International Conference on Energy, Electrical and Power Engineering (CEEPE 2018)

Journal of Physics: Conference Series Volume 1072

Seoul, South Korea
15 – 18 June 2018

ISBN: 978-1-5108-7201-1
ISSN: 1742-6588

Printed from e-media with permission by:

Curran Associates, Inc.
57 Morehouse Lane
Red Hook, NY 12571

CURRAN ASSOCIATES INC. proceedings.com

Some format issues inherent in the e-media version may also appear in this print version.

Printed by Curran Associates, Inc. (2018)

For permission requests, please contact the Institute of Physics
at the address below.

Institute of Physics
Dirac House, Temple Back
Bristol BS1 6BE UK

Phone: 44 1 17 929 7481
Fax: 44 1 17 920 0979

techtracking@iop.org

Additional copies of this publication are available from:

Curran Associates, Inc.
57 Morehouse Lane
Red Hook, NY 12571 USA
Phone: 845-758-0400
Fax: 845-758-2633
Email: curran@proceedings.com
Web: www.proceedings.com

Table of contents

Volume 1072

2018

2018 International Conference on Energy, Electrical and Power Engineering
15–18 June 2018, Seoul, South Korea

Accepted papers received: 27 July 2018
Published online: 13 September 2018

Preface

2018 International Conference on Energy, Electrical and Power Engineering

Conference Committee

Peer review statement

Papers

Information Electronics and Applied Electronics

Detection Method of HVDC Commutation Failure Using Transient Characteristics of Valve Current.....1

Guangqin Chen and Yongjun Zhang

Potential Evaluation of Vertical Axis Hydrokinetic Turbine Implementation in Equatorial River.....7

E. A. Azrulhisham, Z. Z. Jamaluddin, M. A. Azri and S. B. Mohd Yusoff

Real-time Clear-sky Model and Cloud Cover for Direct Normal Irradiance Prediction.....13

Xin Zhao, Haikun Wei, Yu Shen and Kanjian Zhang

A Data-driven Clear Sky Model for Direct Normal Irradiance.....20

Yu Shen, Haikun Wei, Tingting Zhu, Xin Zhao and Kanjian Zhang

Predicting residential electricity consumption using neural networks: *A case study*.....27

Sang Guun Yoo and Hernández-Álvarez Myriam

THE Simulation Study of DC Grounding Electrode Based on CDEGS and ANSYS.....40

Chenglian Ma, Shujian Zhao, Li Sun, Letian Wang and Guangming Zhang

THE Study on the Polar Potential Distribution of DC Grounding in Earth Structures.....45

Chenglian Ma, Shujian Zhao, Li Sun, Letian Wang and Tao Wang

Power Electronics Technology and Applications

Study on the Control System of the Phase Controlled Rectifier for Mine.....51

Quanzhu Zhang, Haiyu Chen, Yuanhong Li and Yonghong Deng

Research of Intelligent Lead-acid Batteries Charger in Coal Mine.....60

Yonghong Deng, Zhishan Liang, Haiyu Chen, Yuanhong Li and Quanzhu Zhang

Discussion on the Over-frequency Generator Tripping Scheme of the Power Grid.....67

Yiting Guo, Haipeng Nan, Xin Guan and Luochang Wu

A New Principle of HVDC Protection Based on Two-terminal Electrical Quantities.....75

Xinyue Zhang, Xiaoning Kang and Yagang Zhang

Complexity Evaluation Model for Smart Grid Management Information System Based on Entropy Theory.....85

Yifu Mo and Yongjun Zhang

Optimization Strategy for Day-ahead Scheduling of Micro-grid Based on the Cost-effective of User Side.....91

Yang Cui, Zifeng Li, Wuzhi Zhong, Zhiwen Yang, Yuting Zhao and Yu Qu

Coordinated Wind Power Accommodating Dispatch Model Based on Combined Electricity and Heat System.....97

Yan Zhuang, Yu Qu, Wuzhi Zhong and Gangui Yan

Investigation of MMC-HVDC Operation Performance Based on Improved Control Strategy.....105

Li Sun, Feng Hu and Sitong Liu

Performance Research on a Twin Screw Expander Applied to a Power Generation System for Recovering Waste Pressure Energy of Natural Gas.....112

S Yao, Y Zhang and H Yang

Genetic Algorithm Based Economic Optimization of Microgrid Including Multi-Energy Supply System.....117

Shujian Zhao, Chenglian Ma, Li Sun, Yuanzhu Meng, Wei Wu and Luheng Huang

New energy development and energy engineering

Heat Transfer Analysis of Blast Furnace Stave Based on Entransy Theory.....122

X Xu and L J Wu

Health, Environmental Effects, the Control of Emission from Power Plants and the Need for a New Emission Capture Technology......129

D K De, U I Ikono and S O Akinmeji

The Research of Theoretical Approaches to Increasing the Reliability of Automated Control Systems in the Energy Sector.....137

Kyaw Zin Lin, Aung Myo Kyaw, Portnov Evgeni Mikalovic and Kyaw Zaw Ye

Comparison of the Multi-Directional Delivery Efficiency of Low-Tonnage LNG and Pipeline Gas in Russia.....147

Pavel Tcvetkov, Daniil Pritulyak and Dmitriy Tananykhin

Analysis and Recommendations of Sand Consolidation Methods to Limit Sand Production in Gas Wells.....154

Dmitry Tananykhin, Pavel Tcvetkov and Vladislav Kamoza

IOP Conf. Series: Journal of Physics: Conf. Series **1072** (2018) 011001 doi:10.1088/1742-6596/1072/1/011001

Preface

It is our great pleasure to welcome you to 2018 International Conference on Energy, Electrical and Power Engineering (CEEPE 2018) held in Seoul, South Korea, June 15-18, 2018. CEEPE2018 is dedicated to issues related to the areas of Energy, Electrical and Power Engineering.

The major goal and features of the conference are to bring academic scientists, engineers, industry and researchers together to exchange and share their experiences and research results, discuss the practical challenges encountered and the solutions adopted. Professors from various areas were invited to deliver keynote speeches providing latest findings in their respective areas of expertise. It was a golden opportunity for the students, researchers and engineers to interact with the experts and specialists to get their advice or consultation.

This proceeding presents a selection from the papers submitted to the conference from universities, research institutes and industries. The papers were selected depending on their quality and their relevancy to the conference. The volume tends to present to the readers the recent advances in the field of Energy, Electrical and Power Engineering and various related areas, such as Advanced Power Semiconductors, Energy Systems, Electrical Machines, Power Electronics and Industry Applications, etc.

Due to the constraint of the capacity and suitability of the topics, 22 submissions were accepted as full papers for publication and presentation in CEEPE 2018. We would like to express our gratitude and appreciation to the authors for their contributions. Many thanks go as well to all of the reviewers who helped us maintain the high quality of manuscripts included in the Proceedings published by Journal of Physics: Conference Series (JPCS). We also express our sincere thanks to the members of the organizing team for their hard work. We wish that all the authors and delegates find CEEPE 2018 interesting, exciting and inspiring. We also wish that our guests will enjoy their stay in Seoul at this beautiful time of the year.

Prof. Om Malik, University of Calgary, Canada
On behalf of CEEPE2018 Organizing Committee
June 15-18, 2018

IOP Conf. Series: Journal of Physics: Conf. Series **1072** (2018) 011002 doi:10.1088/1742-6596/1072/1/011002

Conference Committee

General Chairs

Prof. Omar S. Es-Said, Loyola Marymount University, USA
Prof. Om Malik, University of Calgary, Canada

Program Chairs

Prof. Dennis K. Lieu, University of California, USA
Prof. C. A. Huang, Chang Gung University, Taiwan
Assoc. Prof. Ken Mao, Warwick University, UK

Publicity Chair

Prof. Mohammad Sukri Mustapa, Universiti Tun Hussein Onn Malaysia, Malaysia

International Technical Committees

Prof. Gui-Ping Zhu, Nanjing University of Aeronautics and Astronautics, China
Prof. Pavel Tcvetkov, Saint-Petersburg Mining University, Russia
Prof. Tai-Shen Hung, Chaoyang University of Science and Technology, Taiwan
Prof. Chien-Min Cheng, Southern Taiwan University of Science and Technology, Taiwan
Prof. Cherng-Yuan Lin, National Taiwan Ocean University, Taiwan
Prof. Kew-Yu Chen, Feng Chia University, Taiwan
Prof. Wen-jauh CHEN, National Yunlin University of Science and Technology, Taiwan
Prof. Apirat Siritaratiwat, Khon Kaen University, Thailand
Prof. Noorhana Yahya,Universiti Teknologi PETRONAS, Malaysia
Prof. Sroisiri Thaweboon, Mahidol University, Thailand
Prof. Aziz Ettahir,University Mohammed V of Rabat - Morocco
Prof. Esah Hamzah, Universiti Teknologi Malaysia, Malaysia
Prof. Uma Maheshwera Reddy Paturi,CVR College of Engineering, India
Prof. Kalyan Kumar Ray, Indian Institute of Technology, India
Assoc. Prof. Boonyanit Thaweboon, Mahidol University, Thailand
Assoc. Prof. Zheng Lu, Tongji Universiy, China
Assoc. Prof. Yu-Che Huang, Chaoyang University of Science and Technology, Taiwan
Assoc. Prof. Shu-huei HSIEH, National Formosa University, Taiwan
Assoc. Prof. Willyanto Anggono, Petra Christian University, Indonesia
Assoc. Prof. Kevin Kam Fung Yuen, Singapore University of Social Sciences, Singapore
Assoc. Prof. Huu Loc Nguyen, Ho Chi Minh City University of Technology, Viet Nam
Assoc. Prof. Abu Bakar, Universiti Teknologi Malaysia, Malaysia
Assoc. Prof. Choon Aun Ng, Universiti Tunku Abdul Rahman, Malaysia
Assoc. Prof. Debdulal Das, Indian Institute of Engineering Science and Technology, India
Assoc. Prof. Chong Fai Kait, Universiti Teknologi PETRONAS, Malaysia
Assoc. Prof. Saiful Amri Mazlan, Universiti Teknologi Malaysia, Malaysia
Assoc. Prof. BASHAR S. Mohammed, Universiti Teknologi PETRONAS, Malaysia
Assoc. Prof. Dorin Drignei, Oakland University, USA

Peer review statement

All papers published in this volume of *Journal of Physics: Conference Series* have been peer reviewed through processes administered by the proceedings Editors. Reviews were conducted by expert referees to the professional and scientific standards expected of a proceedings journal published by IOP Publishing.

IOP Conf. Series: Journal of Physics: Conf. Series **1072** (2018) 012001 doi:10.1088/1742-6596/1072/1/012001

Detection Method of HVDC Commutation Failure Using Transient Characteristics of Valve Current

Guangqin Chen, Yongjun Zhang

School of Electric, South China University of Technology, Guangzhou 510641, China

827224094@qq.com

Abstract. In this study, a detection method of high voltage direct current (HVDC) commutation failure using transient characteristics of valve current to solve the problem of no reliable detection method for HVDC commutation failure in electromagnetic transient simulation is designed. By analyzing the transient characteristics of the valve current in the commutation failure process, the commutation failure criterion is extracted and the detection method has been designed. The detection method includes three parts: detection of valve current extreme point and change rate on its both sides, detection of maximum valve current, and detection of continuous non-zero time of valve current after extreme point. Using the PSCAD/EMTDC electromagnetic transient simulation software, a detection method based on the CIGRE BENCHMARCK HVDC model is built up, and the validity and accuracy of the detection method by simulation is verified.

1. Introduction

Commutation failure is one of the most common fault of HVDC transmission systems [1-3], which may reduce equipment life cycle, increases operation and maintenance costs [4], and threatens system operation safety and stability. Fast and accurate detection of commutation failure is an important condition to ensure the safe and stable operation of the system [5]. At present, the criterion and detection of the commutation failure are mainly divided into two types: the measured and the predicted [6]. In general, the measured type of criterion and detection of commutation failure will be more accurate [7]. But based on the comparison of the measured shut-off angle and the limit shut-off angle, due to the existence of a certain blind area, there will be a problem of misjudgment [8]. Aiming at the nature of commutation failure, a commutation failure criterion based on valve current characteristics is proposed in this paper, and a commutation failure detection method using valve current transient characteristics is designed. Based on the model of CIGRE BENCHMARCK HVDC, the PSCAD/EMTDC electromagnetic transient simulation software was used to build a commutation failure detection module and simulations were performed. The simulation results show that the method can accurately and quickly detect whether the system commutation fails.

2. Valve Current Characteristic Analysis During HVDC Commutation Failure Occurs

When the two valves of the inverter commutate, if the commutation process fails to complete, or if the valve is planned to be shut down and the blocking capability cannot be restored during the reverse voltage withstand, it will switch back on when the withstand voltage turns positive, that is, the reverse commutation occurs. At the same time, the valve expected to conduct shuts off again, which is called commutation failure [1, 9-11].

As we know, during the commutation and reverse commutation, the valve current is reduced to zero and maintained for a short period of time (that is, shut-off angle γ)[12]. In severe cases, it even doesn't fall to zero. But when the valve starts to withstand positive voltage, that is, the valve is turned on again or when the next valve starts to withstand the reverse voltage, the commutation process from the valve to the next valve does not end, causing the next valve to start blocking and the current on the first valve to increase again. When the valve connected to the same phase in another half bridge is turned on, due to the short circuit on the DC side, the increase amplitude and increase rate of the current on the valve is increased again.

No matter in simulations or actual projects, the inverter trigger angle is normally 140° during normal operation, while the commutation angle is about 25°, and the turn-off angle is about 15°. Once the fault causes the DC current to increase, the commutation angle will increase substantially at the same time, and the shut-off angle will be reduced. Once it is smaller than the angle corresponding to the minimum time for the valve to recover the blocking capacity, ie the minimum shut-off angle, the commutation will fail. The minimum turn-off angle in real projects is about 7°, which is equivalent to only 0.39 ms, as shown in figure 1(a). In addition, in most cases, reverse commutation occurs when the voltage crossing zero occurs and the commutation process is not completed, as shown in figure 1(b).

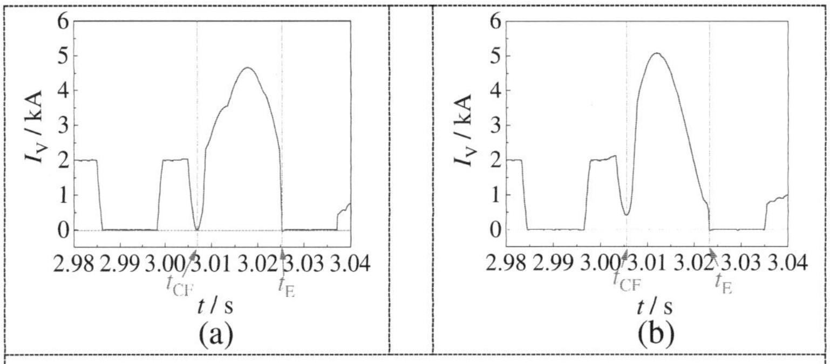

Figure 1. Valve current during commutation failure occurs.

3. Commutation Failure Detection Method Using Valve Current Characteristics

3.1. Commutation Failure Criterion

In summary, the essential process of commutation failure is the reverse commutation. It is very simple and accurate to judge whether the system commutation fails through the transient characteristics of the valve current waveform such as the valve current amplitude and its change rate, or whether there is an extreme point. Therefore, the valve current can essentially reflect whether the system commutation fails, and because of its outstanding features of commutation failure, it is very suitable for the detection of commutation failure. Therefore, based on the transient characteristics of the valve current at the time of commutation failure, a criterion for commutation failure is proposed:

- The valve current waveform has a minimum point, and the rate of decay or increase before or after the minimum value reaches a certain value;
- The maximum value of the valve current after the minimum point is greater than the normal operating current;
- The non-zero duration of the valve current after the minimum point is greater than the normal continuous conduction time of one of the converter valves.

If the above three conditions are satisfied at the same time, it is determined that commutation failure has occurred.

3.2. Commutation Failure Detection Method

(1) Detection of valve current extreme point and change rate on its both sides

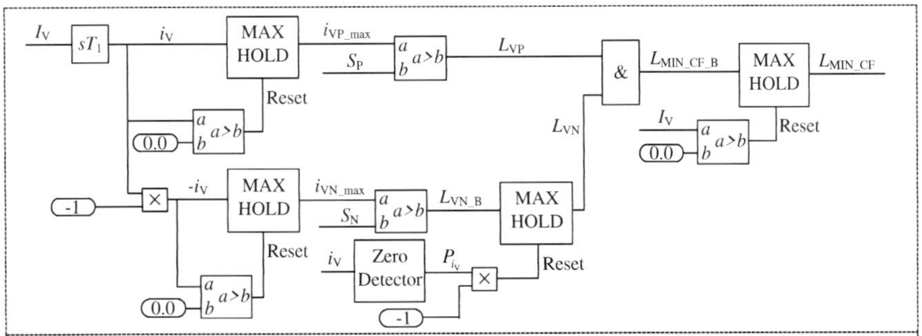

Figure 2. Detection logic of valve current extreme point and the rate of valve current change before or after the extreme point.

Mainly using the derivative of the valve current, the logic of figure 2 is to detect whether there is an extreme valve current, and to judge whether the maximum decay rate before reaching the minimum point is greater than S_N and the maximum increase rate after the minimum point is greater than S_P. When the output signal L_{MIN_CF} is high level 1, there is an extreme point, otherwise it does not exist.

(2)Detection maximum valve current

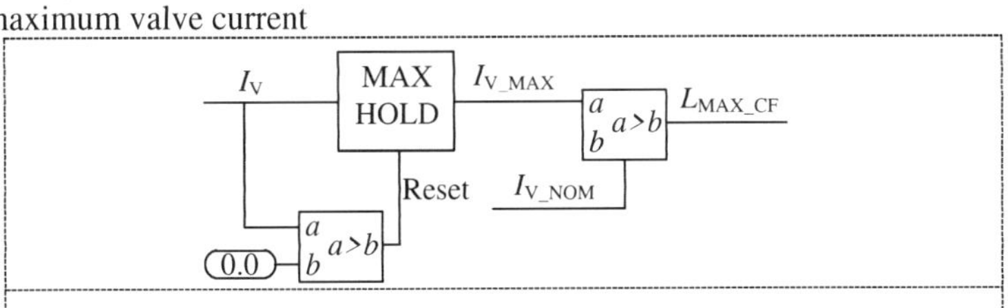

Figure 3. Detection logic of maximum valve current.

The logic of figure 3 is to check whether the maximum value of the valve current is greater than the maximum current I_{V_NOM} when the converter valve is normally turned on. When the output signal L_{MAX_CF} is high level 1, there is over current, otherwise it does not exist.

(3)Detection of continuous non-zero time of valve current after extreme point

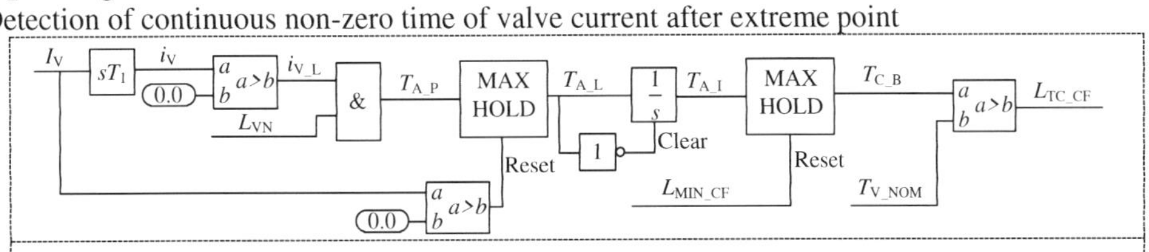

Figure 4. Detection logic of the time that valve current lasts for non-zero after the extreme point.

The logic of figure 4 is to measure the time that valve current continues for non-zero after the extreme point, and it is judged whether the time is greater than the normal time T_{V_NOM}. When the output signal L_{TC_CF} is high level 1, it indicates that the valve current duration after the extreme point meets the commutation failure criterion, otherwise commutation failure does not occur.

(4) Commutation failure detection

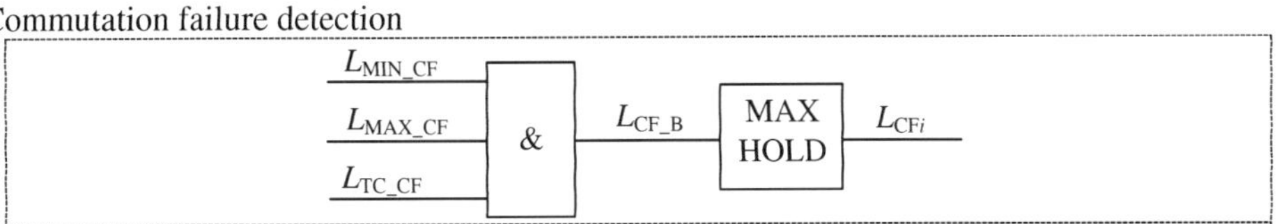

Figure 5. Commutation failure detection.

As shown in figure 5, the signals output by the above three condition detection logics are output by the AND logic and are retained by the MAX HOLD as the final *i*-th valve current commutation failure detection output signal for the 12-pulse converter. *i*=D1, D2,..., D6, Y1, Y2,..., Y6. If there is a valve

IOP Conf. Series: Journal of Physics: Conf. Series **1072** (2018) 012001 doi:10.1088/1742-6596/1072/1/012001

current commutation failure detection output is high level 1, the commutation fails, otherwise commutation failure occurs.

4. Simulation

Based on the CIGRE BENCHMARK HVDC standard test model (shown in figure 6), this paper s up a commutation failure detection module in the electromagnetic transient simulation softw PSCAD/EMTDC and sets different faults on the AC bus of the inverter side to verify the feasibility the proposed commutation failure detection method.

The commutation failure detection module parameter settings are shown in table 1:

Table 1. Parameters of commutation failure detection module.

Parameter	T_1 / (s)	S_N / (kA/s)	S_P / (kA/s)	I_{V_NOM} / (kA)	T_{V_NOM} / (s)
Value	0.001	0.5	0.5	2.2	0.008

To verify the accuracy of proposed commutation failure detection method using the characteristics valve current, the typical three phase faults is selected as an example for analysis.

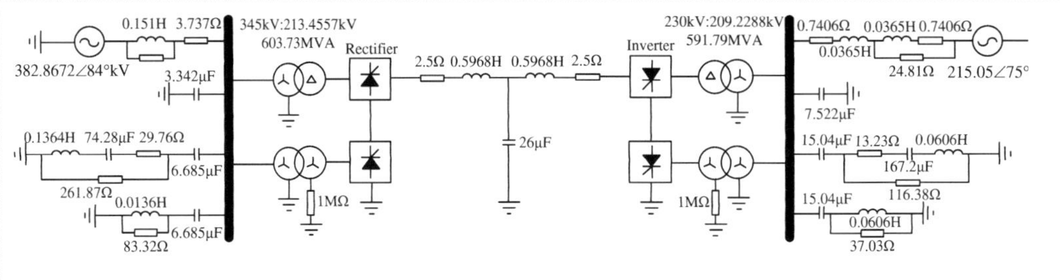

Figure 6. CIGRE HVDC BENCHMARK system circuit diagram.

4.1. Three phase short circuit fault in AC bus at inverter side

The fault starting time is 3.0s, and the duration is 100ms. The short-circuit grounding inductance taken as 0.4H. The simulation results are shown in figure 7.

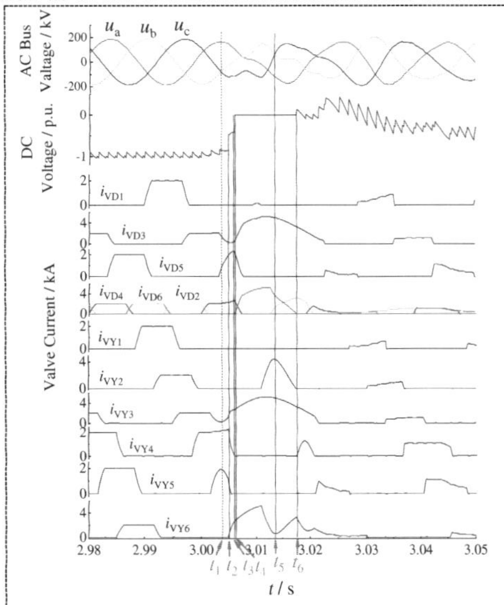

Figure 7. Waveforms of AC voltage at inverter side, DC voltage and valve current under three phase short-circuit fault.

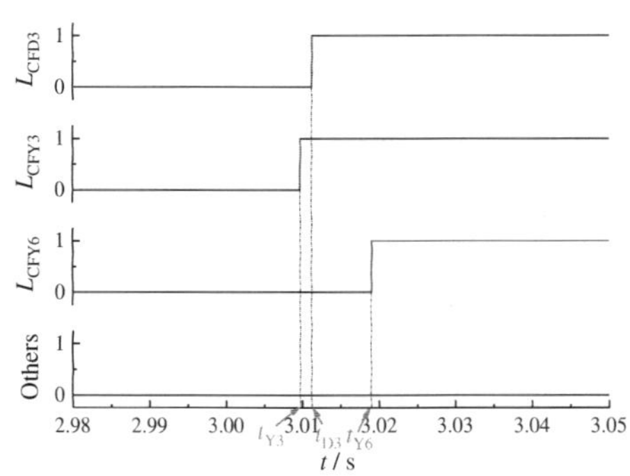

Figure 8. Detection result under three phase short-circuit fault.

As can be seen from figure 7, after three-phase short-circuit fault occurs in the 3.0 s on the inverter side AC bus, the commutation from the valve VY3 to the valve VY5 fails at the first time t_1=3.0035 s, and when the valve VY6 is turned on at t_2=3.0048 s, the inverter side DC voltage drops by about half. At t_3=3.0058s, the commutation failure occurs when the valve VD3 commutates to the valve VD5, and when the valve VD6 is turned on after t_4=3.0077s, the DC side short circuit occurs, so that the DC voltage on the inverter side drops to zero. And it continues until the valve VY6 fails to commutate to the valve VY2 (t_5=3.0138s), that is, the time valve VD6 turns off (t_6=3.0176s).

The commutation failure detection result using the characteristics of valve current is shown in figure 8, and the detected commutation failure order is: valve VY3, valve VD3, and valve VY6. Obviously, it is consistent with the above analysis. In addition, the deviation between the detection result and the switching phase is shown in table 2:

Table 2. Time deviations of detection results during commutation failure occurs under three phase short-circuit fault.

Deviation	t_{Y3}-t_1 / (s)	t_{D3}-t_3 / (s)	t_{Y6}-t_5 / (s)
Value	0.0062	0.0055	0.0052

From table 2, we can see that the reaction time of the proposed commutation failure detection method under three-phase fault on AC bus of the inverter side is only 5 ~ 6ms, indicating that it can quickly provide support for the control and the action of protection system.

Other fault types have also been simulated and the feasibility and effectiveness of the method have been verified. Due to space limitations, further details are not listed.

5. Conclusion

A commutation failure criterion and detection method using valve current characteristics was designed in this paper. This method is simulated by selecting three phase fault on the AC side of the inverter to verify its feasibility. It is proved that this method can effectively, reliably and quickly detect commutation failures. However, it is not limited to three-phase fault condition, and it is still valid under the other fault conditions. It can provide reference and support for commutation failure simulation research work.

6. References

[1] Yongjun Z and Biyun C 2012 The Principle and Application of HVDC Transmission 1st ed. Beijing *Tsinghua University Press* 8

[2] Thio C V, Davies J B and Kent K L 1996 Commutation failures in HVDC transmission systems Discussion *IEEE Transactions on Power Delivery* 11(2) 946-957

[3] Rahimi E, Gole A M, Davies J B, et al 2010 Commutation Failure Analysis in Multi-Infeed HVDC Systems *IEEE Transactions on Power Delivery* 26(1) 378-384

[4] Ha T T, Yongjun Z, Jinbao H, et al 2018 Energy Hub's Structural and Operational Optimization for Minimal Energy Usage Costs in Energy Systems *Energies* 11(4) 707

[5] Hansen A and Havemann H 2002 Decreasing the commutation failure frequency in HVDC transmission systems *IEEE Transactions on Power Delivery* 15(3) 1022-1026

[6] Liang Z, Tao L, Dan Z, et al 2014 Diagnosis of commutation failure in HVDC transmission systems based on voltage or current of converter valve *Power System Technology* 38(12) 3453-3458

[7] Chaoyong H, Xingyuan L, Xiaoming J, et al 2006 Criteria for commutation failure in HVDC transmission system *Power System Technology* 30(22) 19-23 58

[8] Chenhao L, Chang L, Zhenan Z, et al 2017 Analysis of commutation failure in HVDC power transmission system considering prediction control of commutation failure and fault close angle *Automation of Electric Power Systems* 1-9

[9] Son H I and Kim H M 2016 An Algorithm for Effective Mitigation of Commutation Failure in High-Voltage Direct-Current Systems *IEEE Transactions on Power Delivery* 31(4) 1437-1446

[10] Yao S and Yong T 2017 Fast Evaluation of Commutation Failure Risk in Multi-infeed HVDC Systems *IEEE Transactions on Power Systems* PP(99) 1-1

[11] Ruipeng L, Yongli L and Xiaolong C 2017 A control method for suppressing the continuous commutation failure of HVDC transmission *Proceedings of the CSEE* 1-16
[12] Shuyong C, Xinnian L, Jun Y, et al 2005 A method based on the sin-cos components detection mitigates commutation failure in HVDC *Proceedings of the CSEE* (14) 1-6

Potential Evaluation of Vertical Axis Hydrokinetic Turbine Implementation in Equatorial River

E. A. Azrulhisham[a], Z. Z. Jamaluddin[b], M. A. Azri[c], S. B. Mohd Yusoff [d]

[a, b, c] Universiti Kuala Lumpur, Malaysia France Institute, Bandar Baru Bangi, 43650, Malaysia
[d] Faculty of Applied and Creative Arts, Universiti Malaysia Sarawak, 94300, Malaysia

eazisham@gmail.com; zahrul@unikl.edu.my; azrihata@gmail.com; mysaiful@unimas.my

Abstract. In rural areas, small villages are often disconnected from the electrical grid because of lack of economic resources. Long extensions of the grid, only contributing small rural areas with electricity, are not economically beneficial. The temporary solution for the disconnection is to contribute with diesel-generators. These have become common for remote electricity distribution because of their quick installation and mobility. However, the high expenses of the fuel for the diesel generators, along with the challenging fuel transportation because of the scarce to non-existing infrastructure, are not a sustainable solution for electricity distribution in the rural areas. Hydrokinetic turbine is an attractive option for harnessing energy from natural flow of rivers where no static head exists. However, equatorial river system is subjected to high amount of sediment and flowing objects due to the large quantity of vegetation in the area. This leaves the turbine exposed to a higher degree of erosion on the blades and a higher risk of getting clogged. A field study was conducted to attain the velocity and characteristics of sediment in the Niah River located in Miri of Sarawak state, Malaysia. The collected data was analyzed and used as input values during simulations of a turbine model by the Computational Fluid Dynamics. Areas on the turbine exposed to the water with a high velocity and containing a high concentration of sediment were examined. The investigation states that areas on the blade's outer sides are most exposed to erosion and will be in need of protection in form of a harder surface. Adaptation of hydrokinetic turbine for the conditions prevailing in the equatorial river is essential to improve the efficiency and lifetime of the turbine.

Keywords: hydrokinetic turbine, equatorial river, sediment, erosion

1. Introduction

Electrification is generally seen as one of the key indicators that can reduce poverty. Thus, rural electrification is an essential element in bringing about the social and economic development of the underprivileged rural populations. Nevertheless, considering the high cost of distribution and associated transmission loss, grid power supply in rural areas is not economically viable. For most of rural community in Malaysia, the source of power comes from a diesel generator, which operates for about 2 to 3 hours a day. However, the cost of running the generator is relatively expensive as the price of diesel as well as the cost of bringing the fuel to the longhouse can be very high. The off-grid electricity, which can be generated by solar, wind or hydro technology, provides the opportunity to expand the capacity of rural electrification and has distinct advantages for the community as a cost-effective strategy and reliable source of energy.

Photovoltaic solar energy conversion systems have been used to supply electricity for isolated locations in equatorial region. However, the system installed in this area is found to be seriously hampered by tropical weather-related problem. The intense heat and humidity in this region corrode and crack electrical connections and electronic components and encourage mold growth on panels. Tropical climates characterised by fast-changing irradiance conditions due to high variations in cloud coverage influenced the battery charging operation of small solar energy systems [1]. Frequent rains and lightning strikes are also a common cause of system failures in which faulty systems are usually abandoned due to scarcity of spare parts as well as expert know-how among the local community to exert simple maintenance [2].

Wind energy conversion systems have had little applications in remote locations of Malaysia due to poor wind resources. However, a few small turbines, such as those in Sabah and Sarawak are being used in hybrid systems, integrated with solar panels or diesel generators. Similar with the solar energy, strikes by lightning is also one of the major barriers to using wind turbines in equatorial region.

Hydrokinetic turbine is a low pressure run-of-the-river ultra-low-head turbine that will operate on the equivalent of less than 0.2 m of head [3]. The turbines are used to take advantage of the energy in rivers, tidal and ocean currents. For implementations in rivers, hydrokinetic turbines are an attractive alternative for small scale hydropower where the environmental effects from dams can be avoided. The turbines are also a cheaper alternative as investment costs are low and less maintenance work is necessary.

Though, smaller particles in the water such as soil sediments will expose the turbine to erosion. The sediment causes damage to exploited segments of the turbine and will result in alteration of blade profile, increased vibration, fatigue damage, inefficient operation and system failure. The effect of erosion is dependent on the characteristics of the particles, flow of the water and concentration of the sediment as well as the operating conditions of the turbine.

Assessing the sediment properties of a stream where a hydrokinetic turbine is planned is an essential prerequisite to evalaute potential hazards for the turbine and thereby contribute to a higher accessibility to electricity in remote areas. The objective of the current study is to investigate the areas on the turbine most exposed to erosion as of the high amount of sediment in the water and to improve knowledge of turbine wear with regards to relevant suspended sediment parameters.

2. Methodology

Data were collected during the field study in Niah River located in Miri of Sarawak state, Malaysia. Niah River is an equatorial stream located about 3 km from the small town of Batu Niah, 110 km south-west of Miri on the island of Borneo. The 105 km river originates from the Niah basin with total area of 1,343 km^2. The area has a tropical rainforest climate with two monsoon seasons: the southwest monsoon, which is the dry season from April to September, and the northeast monsoon, which is the wet season from October to March. The annual rainfall is around 250 to 380 cm with the air temperature between 23 °C to 32 °C the whole year round. The morphological cross-section of River Niah is shown in Fig. 1.

Fig. 1. Morphological cross-section of River Niah

2.1. Velocity of flowing water

Sediment erosion is caused by the stroke of sediment in the flowing water making an impact on the solid surface. The amount of sediments in the water is highly connected to the current amount of discharge and the current water velocity. Measurements were carried out during the field study to establish the velocity of the water in the Niah River. The velocity of the flowing water in the river was measured by using a Swoffer propeller type current meter consisting of moving blades that rotates with the water velocity. The device had the possibility to be used at different depths and was in this case used to measure the velocity at depth level of 0.4 m, 1.5 m and 3.0 m. The velocity of the water varies considerable depending on where in the river the measurements are taking place. The water velocities measured at the several subsections as shown in Figure 6. For each measuring point, 3 to 5 readings were taken and averaged to give a mean point velocity to reduce the error due to variation in water flow.

2.2. Sediment profiling

To collect information about the sediment, water was sampled from specific depths and afterwards analysed using a Laser In-Situ Scattering and Transmissometry (LISST) instrument shown in Fig. 2 The water was sampled in vertical line at three different depths using a US P-61-A1 point integrating sampler on a moving boat along the whole section of the river. Due to the streamlined designed and the sampling method it can provide undisturbed water samples. This device provides the input samples for the LISST instrument in which laser refractometry is used to measure volumetric concentrations for particles based on 44 logarithmically spaced size classes between 0.34 to 500 μm. In-situ measurements of particle size and concentration is provided by a laser beam directed into the sample volume where particles in suspension scatter, absorb, and reflect the beam. The scattered laser light is received by a series of ring shaped detectors of progressive diameters that allow measurement of the scattering angle of beam. Particle size can be calculated from knowledge of this angle either by using the Frauhofer approximation or the exact Lorenz-Mie solution. Particle size dependency is eliminated by basing concentration measurements on these measured particle sizes, thus allowing both particle size and volume concentration to be simultaneously measured [8].

Fig. 2. Laser In-Situ Scattering and Transmissometry (LISST) instrument

2.3. Sediment erosion

Sediment erosion is caused by the stroke of sediment in the flowing water making an impact on the solid surface. The effect of erosion is dependent on the characteristics of the particles and also the operating conditions of the turbine, such as the flow of the water [4]. The rate of increased effect from erosion from sediment is also increasing with the concentration of the sediment. The actual effect of erosion is very complex because of the many different factors affecting the result [5]. The phenomenon is almost impossible to simulate with the accuracy of the actual flow of the river. However, experimental trials have been conducted to determine the erosive wear on a Pelton turbine buckets [6]. From the trials, a simple equation, determining the amount of erosion per unit of water discharge *W* as a function of current conditions, could be established using Equation (1) in which s represents the silt size, c the sediment concentration, *v* the water velocity and *t* the operation time.

$$W = 4.02 \bullet 10^{-12}(s)^{0.0567}(c)^{1.2267}(v_w)^{3.79}(t) \tag{1}$$

2.4. Computational fluid dynamics simulation

Simulations of the vertical axis hydrokinetic turbine were conducted using computational fluid dynamics (CFD) software tool ANSYS Fluent. The turbine to be implemented was constructed based on 3 rotor blades design with size of 0.85 m wide and 0.70 m height as shown in Fig. 3.

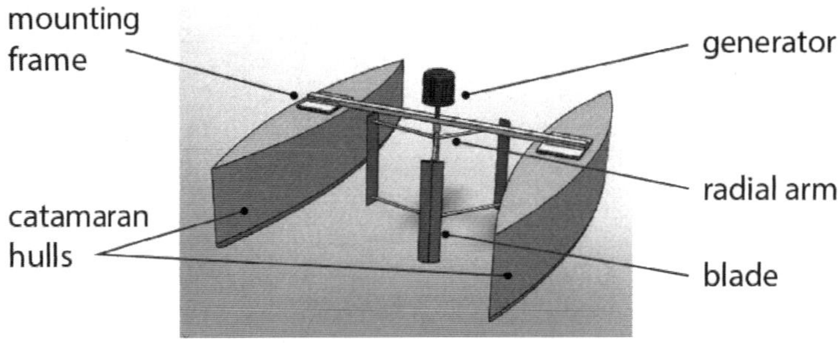

Fig. 3. Vertical axis hydrokinetic turbine

The turbine model was simplified to 2D and the planar surface is set to be symmetric across the direction normal to the plane (z-axis). In the analysis the pressure from water hits the turbine blades and subsequently yields certain magnitude of resultant force that initiate the rotation. The turbine rotation is driven by the moment produced from the net force acting on the blade in radial direction and the arm length. The CFD analysis was used to identify the areas on the turbines which achieve the highest velocity. The geometry for the analysis was prepared in ANSYS Spaceclaim Direct Modeler^{TM} with number of blade of 3 and the NACA profile of 0018. The 2D plane was generated considering the traced profile and the pattern tool was used to create the 3 blades configuration as shown in Fig. 4. The element size of the mesh used during the simulations of the turbine was set to 1mm, containing 104,832 elements and 105,605 nodes. In order to obtain fine results across the blade profile, inflation was introduced and bounded along the profile edge.

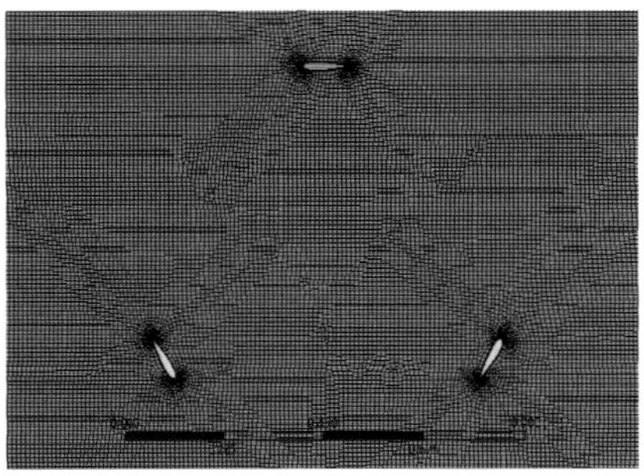

Fig. 4. Vertical axis hydrokinetic turbine

The particles were assumed to have the same velocity as the water, the higher the velocity of the water hitting the turbine, the higher impact on the turbine from particles. The areas on the blades which achieve the highest velocity were determined as the most vulnerable areas as the majority of the erosive effect will occur at these areas.

The information of the collected extent of sediment in the water was used as inlet data of moving particles in fluid flow during simulations of the turbine. The mean concentration of the sediment from the different depths were used to give an applicable result as the depth of implementation was not yet determined. The amount of sediment at the different depths could indicate whether the erosion on the turbine might differ depending on the depth, or not, which should be considered when determine the depth of implementation.

3. Results and Discussion

The velocity of the water varies considerable depending onwhere in the river the measurements are taking place [7]. This was confirmed during the field study. The water velocities measured at the several subsections with different depth of the river show that the flow velocity occurs in the central of main channel region, which decreases towards the side banks as summarized in Table 1. The water level also varies widely in the Niah River depending on seasons. Because of the portability of the turbine it is however easy to move to an area with higher level and velocity during dryer periods. The mean velocity of measurements of the central region of the river was calculated to be 1.02 m/s, and it was used during CFD simulations of the turbine.

Table 1. Velocity of river flowing water at different subsection

	Subsection 1	Subsection 2	Subsection 3	Subsection 4	Subsection 5
Depth 0.4 m	0.19 m/s	0.63 m/s	1.33 m/s	0.52 m/s	0.31 m/s
Depth 1.5 m	0.29 m/s	0.77 m/s	0.97 m/s	0.66 m/s	0.42 m/s
Depth 3.0 m	0.26 m/s	0.64 m/s	0.78 m/s	0.58 m/s	0.29 m/s

The measured suspended sediment data consist of 30 observations from ten locations of the river indicates that all samples are dominated by silt and fine sand particle with the mean size range of < 50 μm. A detailed box-plot representation of the mean size and concentration distribution of suspended sediment particle for depth level is shown in Fig. 5.

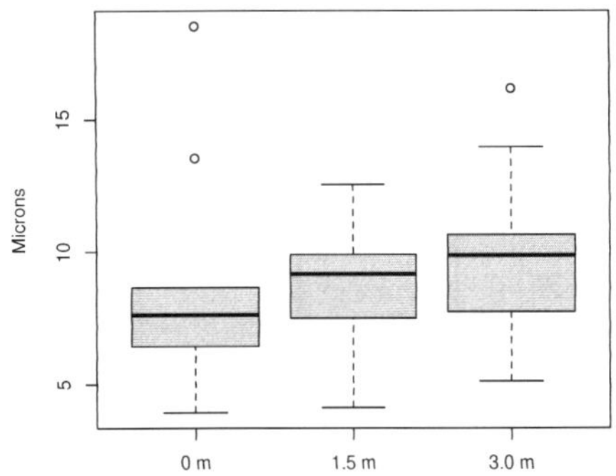

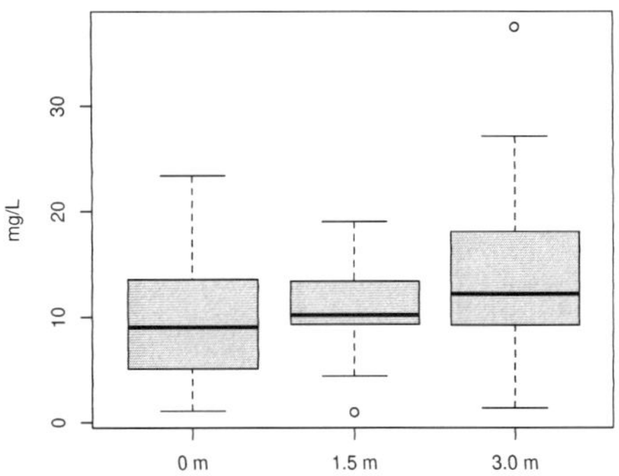

Fig. 5. Mean size and concentration distribution of suspended sediment particle at different depth levels

The erosive wear that may occur at different depth level was established by equation (1) as shown in Table 2.

Table 2. Erosive wear estimation at different depth levels

Depth Level	Mean Sediment Size (μm)	Mean Sediment Concentration (mg/l)	Erosive Wear (mg erosion / l
0 m	8.698	10.019	9.074×10^{-10}
1.5 m	8.866	10.670	9.813×10^{-10}
3.0 m	10.025	14.667	14.601×10^{-10}

The sediment in the water makes the areas that are exposed to the high velocities even more exposed to a higher degree of erosion than other areas of the turbine. Thereby, the velocity is an important factor to detect the most vulnerable areas of the turbine. As erosion is a complex phenomenon and almost impossible to simulate, the focus of CFD analysis has been to determine the vulnerable areas of the turbine, where most sediment will hit. A visual indication as shown in Fig. 6 was used to establish more exactly where the highest water velocity interacts with the turbine blades and the force on each blade in the radial direction.

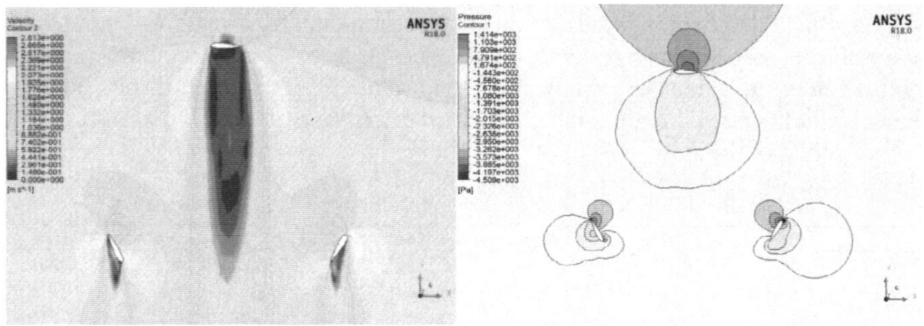

Fig. 6. Areas where highest water velocity interacts with the turbine blades and the force on each blade

The results confirm that outer surface of the blades is exposed to higher velocities, and they are thereby more exposed to the effect of erosion. The increased pressure forcing the blades to move results in a higher velocity exerted on the part of the blade than compared to the connected radial arm. According to simulations, it would be recommended to add a coating on the outside of the blades as these are the areas most exposed to a high velocity and thereby a high concentration of silt.

4. Conclusion

Hydrokinetic turbine is an attractive option for harnessing energy from slow flowing water where no static head exists. However, equatorial river system is subjected to high amount of sediment that will exposed the turbine to erosion. Therefore, ways to protect the turbines were investigated. Measurements were carried out during a field study in Niah River located in Miri of Sarawak state, Malaysia where water velocity as well as size and concentration of suspended sediment were analyzed. A model of the turbine to be implemented was simulated using ANSYS Fluent in which the collected information from the river was used as the input data. Areas of the turbine where sediment-full water hits with a high velocity were detected and considered as the most vulnerable areas for erosive effect. The method is particularly applicable in predicting the silt erosion on solid surface of turbine components due to dynamic action of suspended sediment. In view that the efficiency of the turbine will decreases with the increase in the erosive wear, it is therefore necessary to improve on the hydraulic and protective design of the component.

5. References

[1] Anyi M., Kirke B. Evaluation of small axial flow hydrokinetic turbines for remote communities. *Energy for Sustainable Development*, 2010; (14): 110-116.

[2] Feron S. Sustainability of off-grid photovoltaic systems for rural electrification in developing countries, a review. *Sustainability,* 2016, (8): 1326.

[3] Khan M.J., Bhuyan G., Iqbal M.T., Quaicoe J.E. Hydrokinetic energy conversion systems and assessment of horizontal and vertical axis turbines for river and tidal applications: A technology status review. *Applied Energy*, 2009, (86): 1823-1835.

[4] Neopane H.P., Sujakhu S., Shrestha S., Subedi, K., Basnet A. Study of sediment erosion in hydraulic turbine using rotating disc apparatus, *International Conference on Developments in Renewable Energy Technology (ICDRET)*, 2012: pp. 1-4.

[5] Padhy M.K., Saini R. A review on silt erosion in hydro turbines. *Renewable and Sustainable Energy Reviews*, 2008, 12(7): 1974-1987.

[6] Padhy M., Saini R. Effect of size and concentration of silt particles on erosion of Pelton turbine buckets. *Energy*, 2009, 34(10): 1477-1483.

[7] Sornes K. 2010. Small-scale water current turbines for river applications. *Zero Emission Resource Organisation (ZERO)*, 2010.

Acknowledgements

This paper is part of an ongoing research project "Empirical Sediment Settling Prediction System and Optimization Tools for Small Hydro Power Plant" which is financially supported by the Ministry of Energy, Green Technology and Water Malaysia under the Malaysian Electricity Supply Industries Trust Account (KeTTHA-AAIBE).

IOP Conf. Series: Journal of Physics: Conf. Series **1072** (2018) 012003 doi:10.1088/1742-6596/1072/1/012003

Real-time Clear-sky Model and Cloud Cover for Direct Normal Irradiance Prediction

Xin Zhao, Haikun Wei, Yu Shen, Kanjian Zhang

Key Laboratory of Measurement and Control of CSE, Ministry of Education, School of Automation, Southeast University, Nanjing 210096

E-mail: hkwei@seu.edu.cn

Abstract. Direct normal irradiance (DNI) prediction is of great significance for the concentrated solar power (CSP) generation and grid work. The clouds are the main cause of DNI intensity reduction and drastic changes. Therefore, the real-time clear-sky DNI model coupled with cloud cover is proposed for short-term DNI prediction in this study. The Linke turbidity coefficient is used to develop a clear-sky model, and real-time adjustment of the coefficients is achieved by identifying clear -sky period. Then, the theoretical clear-sky value of current day can be obtained. By the combination of the historical cloud cover and theoretical clear-sky value, Auto Regressive Moving Average (ARMA) and Artificial Neural Network (ANN) are used to develop linear and nonlinear models. Experiment with the data in the National Renewable Energy Laboratory (NREL) database, the simulation results show that this approach using cloud cover and clear-sky information can improve the forecasting accuracy.

1. Introduction

Solar energy is a renewable resource that has the advantages of cleanliness, extensiveness, and potential economics. It plays an important role in long-term energy strategies. Solar power generation is a common way of using solar energy which has become the third renewable energy source in terms of global capacity [1]. Electricity can be generated from solar irradiance by either non-concentrated photovoltaic modules or by concentrated solar power (CSP) [2]. CSP refers to the power generation that converts concentrated sunlight directly into electricity through high conversion efficiency photovoltaic cells. CSP output calculations only use the direct normal irradiance (DNI) component. Hence, accurate CSP output predictions require an estimation of the local DNI [3,4].

DNI is the difference between extraterrestrial irradiance above the atmosphere and atmospheric losses due to absorption and scattering. The losses depend on the length of light through the atmosphere (depending on the sun's elevation), water content, clouds, and other content [5]. Currently, many clear-sky models have been proposed to calculate the clear-sky DNI, when there is no cloud occulting the Sun. Typically, the establishment of the physical model by considering water vapour and aerosol concentrations in the atmosphere can obtain the clear-sky DNI value. Among the clear-sky DNI models, REST2 [6] has proven to assess DNI with high accuracy. The model needs measured value of the band transmittances for Rayleigh scattering, uniformly mixed gases absorption, ozone absorption, nitrogen dioxide absorption, water vapour absorption, and aerosol extinction. However, the measured

IOP Conf. Series: Journal of Physics: Conf. Series **1072** (2018) 012003 doi:10.1088/1742-6596/1072/1/012003

physical quantities are difficult to obtain and not available at any time. The clear-sky empirical model is relatively easy to calculate. The simplest clear-sky DNI models are usually based on the solar zenith angle and empirical correlations derived from experimental measurements. In the literatures [7,8], the clear-sky DNI is obtained from an eight-order polynomial of the cosine of the solar zenith angle. The change of the parameters in this model can't be a good reaction to different atmospheric conditions. It is not easy to figure out the discipline of the parameters change because of too many parameters. Another model based on Linke turbidity coefficient is also widely used [9-11]. Atmospheric conditions are well reflected by atmospheric turbidity which varies greatly over the days. Therefore, real-time adjustment and statistics of the turbidity coefficient combined with the clear-sky time period enable a more accurate DNI prediction.

Under cloudy conditions, the clouds are the main cause of DNI intensity reduction [12,13]. Therefore, DNI has drastic changes and is hard to forecast. The clearness index is a parameter depicting attenuation of DNI in a cloudy day, which is derived from the attenuation of the clear-sky DNI caused by clouds. It changes from 0, when a thick cloud is occulting the Sun, to 1, when there is no cloud occulting the Sun. In literatures [14,15], it validates that there is a certain correlation between cloud cover and clearness index. Thus cloud cover data may be added to modify and estimate DNI under cloudy conditions. Based on this, this paper proposed a short-term DNI forecasting model with real-time clear-sky model and cloud cover. The method first develops real-time adjustment clear-sky model. Then, calculating the theoretical clear-sky value. Finally, the linear ARMA and non-linear ANN model are developed by history cloud cover data and DNI data, respectively. The method aimed to improve DNI forecasting accuracy and the objectives of the present work were: 1) to adjust clear-sky model in real-time by identifying the clear-sky time period; 2) to develop ARMA and ANN model using the cloud cover data and the theoretical clear-sky DNI.

2. Methodology

2.1. Clear-sky model

In 1922, Linke proposed to calculate clear-sky DNI as the product of the optical thickness of a water- and aerosol-free atmosphere (a clear and dry atmosphere) (δ_{cda}) and the Linke turbidity coefficient (T_L) [9]. It further improves the Linke formulation by adding a value of the altitude of the power plant. It is expressed as equation (1).

$$I_{cs} = (0.664 + 0.163 \cdot \exp(h/8000)) \cdot I_0 \cdot \exp(-0.09 \cdot m \cdot (T_L - 1)) \tag{1}$$

where h is the altitude of the power plant, I_0 is the extraterrestrial solar irradiance, m is the air mass. I_0 can be computed using a very simple expression given by Eq. (3):

$$I_0 = (1 + 0.0334\cos(2d\pi/365)) \cdot I_0' \tag{2}$$

where d is the day of year, ranging from 1 (January 1) to 365(December 31). I_0' is the solar constant 1367 Wm^{-2}.

m is modified by Kasten and Young in 1989, defined as

$$m = \frac{1}{\cos(z) + 0.50572 \cdot (96.07995 - z)^{-1.6364}} \tag{3}$$

where z is zenith angle.

So, the atmospheric turbidity coefficient is calculated by inverting equation (4):

$$T_L = 1 + (\ln(\frac{(0.664 + 0.163 \cdot \exp(h/8000)) \cdot I_0}{I_{cs}}) \cdot \frac{1}{0.09 \cdot m}) \tag{4}$$

Figure 1 shows the profiles of DNI and atmospheric turbidity coefficient (T_L) during a clear-sky. As can be seen from the figure, T_L is smaller in the morning and afternoon in the day, and the range of change is about 0.3.

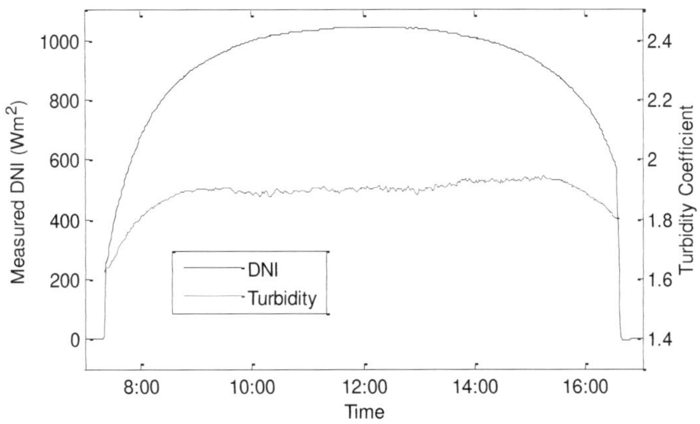

Figure 1. The DNI and atmospheric turbidity coefficient during a clear-sky.

The correlation of clear-sky DNI and T_L is poor when the zenith angle is large. In this paper, the clear-sky model with zenith angle less than 85 degrees is considered. The turbidity coefficient is calculated by the mean of a day, named T_{LD}. Figure 2 shows the comparison of the fitting curve of the clear-sky model and the measured clear-sky curve.

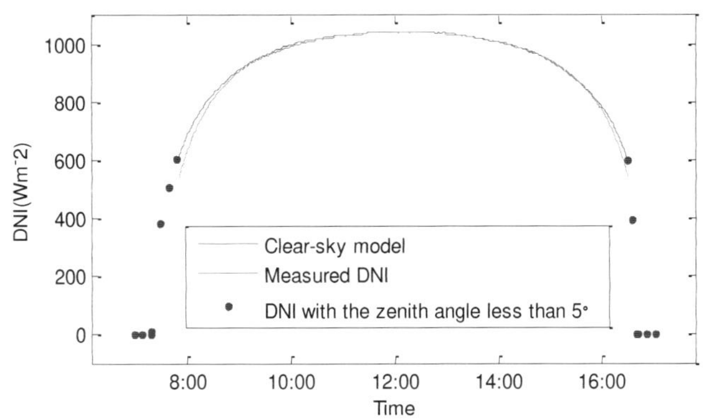

Figure 2. The comparison of clear-sky model and measured DNI curves

2.2. Clear-sky period identifying

Once obtained a clear-sky curve fitted to a clear day, the next step consists of identifying periods with irradiance similar to the clear sky irradiance. Since there are too few data with clear-sky in the whole day, it is very necessary to detect clear-sky period. The T_{LD} obtained from the clear-sky period are then counted. The clear-sky DNI is identified by the combine of two criterion [16]. The first criterion depends on the clearness index, as follows equation (5).

$$D_{cs}^{i} = abs(\frac{I_{cs}^{i} - I_{m}^{i}}{I_{cs}^{i}}) \quad (5)$$

where I_{cs}^{i} is i th clear-sky DNI, I_{m}^{i} is i th measured DNI. I_{cs}^{i} is the determined by the nearest clear day model. The threshold D_{cs}^{i} values correspond to the location. In this study, when $D_{cs}^{i} < 0.25$, it is considered as a clear-sky value.

The second criterion the length of the DNI curve. The points extracted from the first criterion are further adjusted. The points in a continuous interval are used to calculate the curve length. The measured curve length is comparison with the initial model curve length.

$$L_{cs}^{i} = \sum_{j=1}^{t} \sqrt{(I_{cs}^{i-j} - I_{cs}^{i-j-1})^2 + (T^{i-j} - T^{i-j-1})^2}$$

$$L_{m}^{i} = \sum_{j=1}^{t} \sqrt{(I_{m}^{i-j} - I_{m}^{i-j-1})^2 + (T^{i-j} - T^{i-j-1})^2} \tag{6}$$

$$LD_{cs}^{i} = abs(\frac{L_{cs}^{i} - L_{m}^{i}}{L_{cs}^{i}})$$

where I_{cs}^{i-j} is $i-j$ th clear-sky model DNI, I_{m}^{i-j} is $i-j$ th measured DNI, T^{i-j} is $i-j$ th, t is a continuous time period. It is considered as a clear-sky value when LD_{cs}^{i} is less than the threshold 3. Figure 3 shows the identifying clear-sky period under the cloudy conditions and comparison of initial model and the adjusted model.

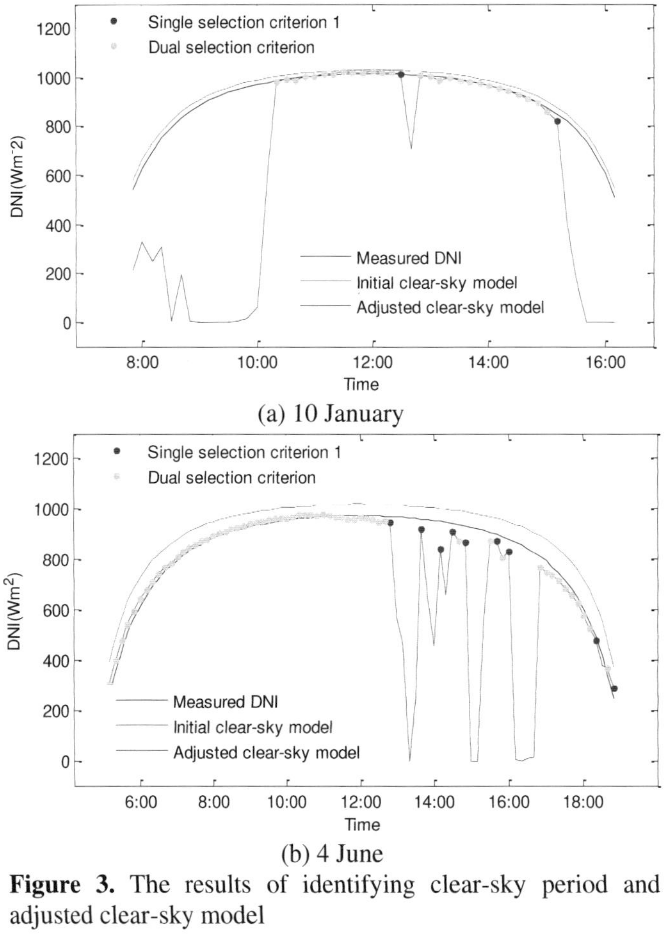

Figure 3. The results of identifying clear-sky period and adjusted clear-sky model

2.3. *DNI forecasting model*

The proposed method explicitly considers the effect of cloud cover on solar irradiance. 19 January 2014 is a full clear-sky which is selected to train the initial model. Then dual criteria are used to identify the clear-sky period for each day. Calculation T_{LD} based on the detected clear-sky period, and the new T_{LD} is used to build the real-time clear-sky model. If there is no clear-sky period at the current day, the clear-sky model in a nearest day is chosen to calculate the theoretical clear-sky value. For testing data, each added point is used to distinguish whether it is clear-sky period, the value of T_{LD} is adjusted in real time. By feeding the cloud cover ahead of 3 moment and theoretical clear-sky value as model inputs, ARMA and ANN were used to develop linear and nonlinear models, respectively.

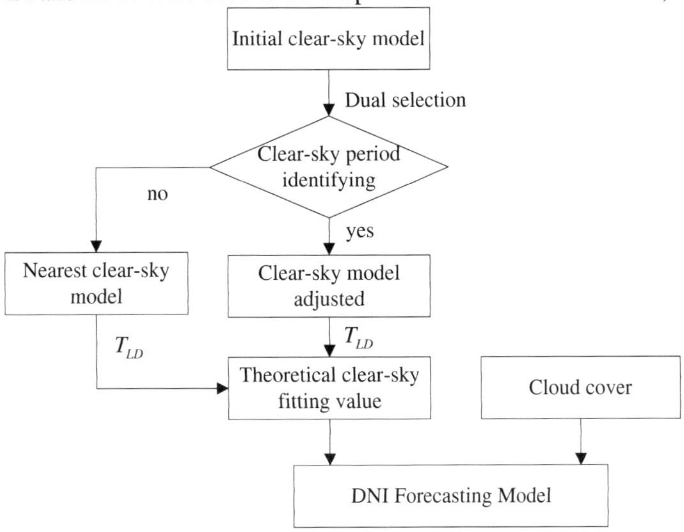

Figure 4. Flow chart of forecasting methods

3. Experimental data and preparation

3.1. DNI and cloud cover dataset DNI forecasting model

In this paper, all the data are from the database of the NREL. The DNI values are measured by a Kipp & Zonen pyrheliometer mounted on a Sci-Tec tracer. The cloud cover are calculated by a Yankee Total Sky Imager Model 880. The instrument captures all sky images, analyzes them, and computes the percentage of cloud cover. The sampling frequency of DNI and cloud cover are selected 10 minute interval. The DNI forecasting model is established with the data from 1 January 2014 to 30 September 2014. The testing data is from 1 October 2014 to 31 December 2014.

3.2. Forecasting performance evaluation

The models are evaluated synthetically using the following evaluation criteria:
1) normalized root mean square error ($nRMSE$)

$$nRMSE = \frac{\sqrt{\frac{1}{N}\sum_{t=1}^{N}\left(y_t - \hat{y}_t\right)^2}}{y_{\max} - y_{\min}} \quad (7)$$

2) mean absolute error (MAE)

$$MAE = \frac{1}{N} \times \sum_{t=1}^{N}\left|y_t - \hat{y}_t\right| \quad (8)$$

where y_t and $\hat{y}_t$ are the measured and predicted DNI at time t, $y_{\max}$ and $y_{\min}$ the maximum and minimum of observed values of DNI, N is the number of test data.

4. Performance analysis

In the present study, all programs were implemented in MATLAB software. In order to verify the validity of the proposed algorithm, three forecasting models were constructed: Persistence model, ARMA, and ANN. The persistence model uses the current value as the prediction, i.e., at time t, the prediction $\hat{y}_{t+1} = y_t$. For the last two models, the model inputs are calculated theoretical clear-sky DNI and cloud cover ahead of 3 moment. The prediction results are shown in Table 1.

Table 1. The DNI forecasting results with different methods.

	Clear-sky model		*Real-time clear-sky*	
Model	*nRMSE*	*MAE*	*nRMSE*	*MAE*
Persistence	15.79	79.02	/	/
ARMA	15.17	80.72	14.93	79.10
ANN	14.72	77.13	14.21	75.45

From Table 1, it can be seen that the ANN coupled with real-time clear-sky model can obtain the best prediction compared with other model predictions. The effect of the nonlinear model ANN is better than that of the linear model ARMA. The clear-sky model, which is corrected by real-time, is more favorable for DNI prediction.

Figure 5 shows the partial forecasting results in different day. It can be seen from the Figure 5(a) and Figure 5(c) that the prediction of the real-time clear sky model with ANN is better under the most clear-sky conditions. The results by coupling real-time clear sky model with ANN is relatively gentle. For the overcast day, the prediction is relatively poor shown in Figure 5(b) and Figure 5(d). Cloud has more complex effects on irradiation. Due to adoption of cloud cover as model input, it can only reflect part of the cloud features. The improvement of the overall prediction is not obvious. The proposed method is more suitable for the most clear-sky day.

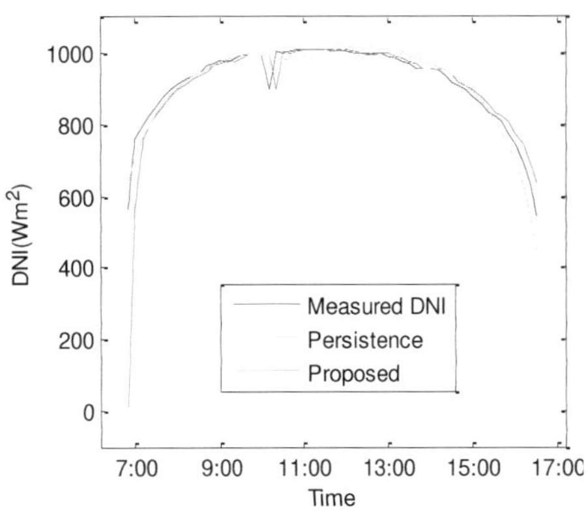

(a) 13 October

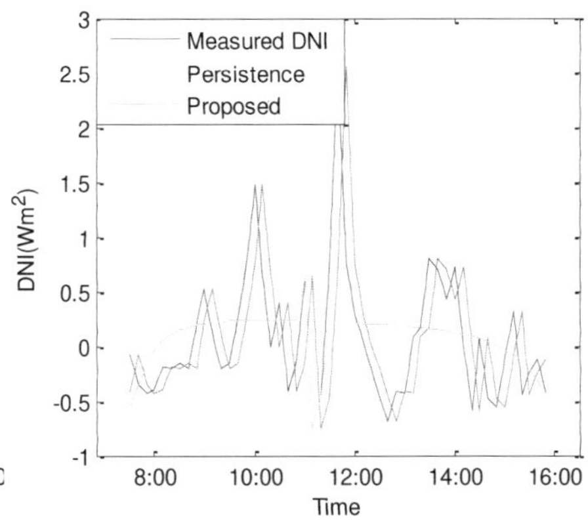

(b) 15 November

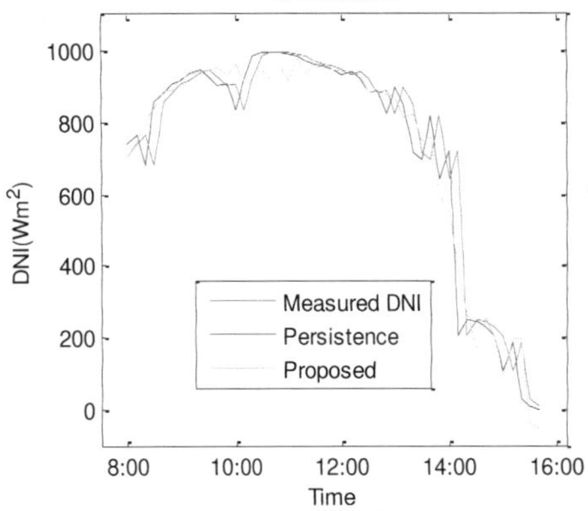

(c) 12 December

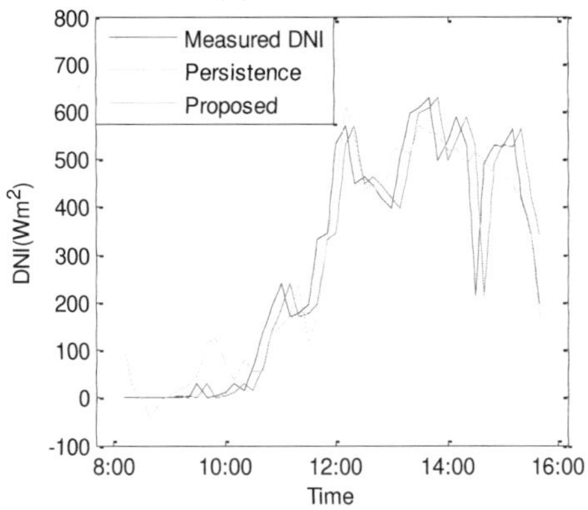

(d) 30 December

Figure 5. The comparison of forecasting results in different day

5. Conclusion

The DNI forecasting method coupled the clear-sky model and cloud cover was proposed in this study. In order to modify the clear-sky model, the real-time clear-sky model was adjusted by identifying the clear-sky period. The modified clear-sky model was for calculating theoretical clear-sky values. Then cloud cover data was added to estimate DNI under cloudy conditions. In order to verify the proposed method, the DNI and cloud cover data from NREL database were used. Coupled with ARMA and ANN model respectively, the results showed that the prediction of the real-time clear-sky adjustment is better. For DNI prediction, the nonlinear model is superior to the linear model. Because cloud cover can only reflect part of the cloud features, more useful cloud features extraction become further research.

6. References

[1] Barbieri F, Rajakaruna S and Ghosh A 2016 Very short-term photovoltaic power forecasting with cloud modeling: A review *Renewable & Sustainable Energy Reviews* **75**.
[2] Zhu T, Wei H, Zhao X, Zhang C, Zhang K 2017 Clear-sky model for wavelet forecast of direct normal irradiance *Renewable Energy* **104** pp 1-8.
[3] Law EW, Prasad AA, Kay M and Taylor RA 2014 Direct normal irradiance forecasting and its application to concentrated solar thermal output forecasting – A review *Solar Energy* **108** pp 287-307.
[4] Nou J, Chauvin R, Thil S and Grieu S 2016 A new approach to the real-time assessment of the clear-sky direct normal irradiance *Applied Mathematical Modelling* **40** pp 7245-64.
[5] Bone V, Pidgeon J, Kearney M and Veeraragavan A 2018 Intra-hour direct normal irradiance forecasting through adaptive clear-sky modelling and cloud tracking *Solar Energy* **159** pp 852-67.
[6] Gueymard CA 2008 REST2: High-performance solar radiation model for cloudless-sky irradiance, illuminance, and photosynthetically active radiation – Validation with a benchmark dataset *Solar Energy* **82** pp 272-85.
[7] Chu Y, Pedro HT and Coimbra CF 2013 Hybrid intra-hour DNI forecasts with sky image processing enhanced by stochastic learning. *Solar Energy* **98** pp 592–603.
[8] Quesada-Ruiz S, Chu Y, Tovar-Pescador J, Pedro HTC and Coimbra CFM 2014 Cloud-tracking methodology for intra-hour DNI forecasting *Solar Energy* **102** pp 267–275.
[9] Linke, F 1922 Transmission-Koefficient und Tru¨ bungsfaktor *Beitr. Phys. fr. Atmos.* **10** pp 91–103.
[10] Ineichen P, Perez R 2002 A new airmass independent formulation for the Linke turbidity coefficient *Solar Energy* **73** pp 151-7.
[11] Gueymard CA, Ruiz-Arias JA 2015 Validation of direct normal irradiance predictions under arid conditions: A review of radiative models and their turbidity-dependent performance *Renewable & Sustainable Energy Reviews* **45** pp 379-96.
[12] Engerer NA, Mills FP 2014 KPV: A clear-sky index for photovoltaics *Solar Energy* **105** pp 679-93.
[13] Luo L, Hamilton D, Han B 2010 Estimation of total cloud cover from solar radiation observations at Lake Rotorua, New Zealand *Solar Energy* **84** pp 501-6.
[14] Myers DR 2007 General cloud cover modifier for clear sky solar radiation models *Proc Spie* **6652**.
[15] Yang D, Jirutitijaroen P and Walsh WM 2012 Hourly solar irradiance time series forecasting using cloud cover index *Solar Energy* **86** pp 3531-43.
[16] Larrañeta M, Reno MJ, Lillo-Bravo I and Silva-Pérez MA 2017 Identifying periods of clear sky direct normal irradiance *Renewable Energy* **113**.

Acknowledgments

This work was supported by the National Natural Science Foundation of China (Grant No. 61374006, Grant No. 61773118 and Grant No. 61703100), and by the Fundamental Research Funds for the Central Universities.

A Data-driven Clear Sky Model for Direct Normal Irradiance

Yu Shen, Haikun Wei, Tingting Zhu, Xin Zhao and Kanjian Zhang

Key Laboratory of Measurement and Control of CSE, Ministry of Education, School of Automation, Southeast University, Nanjing 210096, China

Email: hkwei@seu.edu.cn

Abstract. Solar energy is an ideal new energy source for power systems. In order to integrate solar energy into the power grid, an evaluation of the irradiance input to solar power systems is required in many applications. Clear-sky models describe the maximum input of solar power systems, which are particularly important for solar irradiance forecasting and numerical weather prediction. The existing clear sky models are empirical or semi-empirical models. In order to make use of the information in historical big data, a data-driven clear sky model for direct normal irradiance is proposed in this paper. Firstly, a clear sky detection algorithm is designed to collect clear-sky data from historical big data automatically. After that, the weighted k-Nearest Neighbors regression algorithm is used to calculate clear-sky direct normal irradiance by historical clear-sky data. The database of the National Renewable Energy Laboratory (NREL) is used for experiments to evaluate the performance of the proposed model, compared with other three clear sky models. The results of the experiments show that the proposed model is more accurate than other three clear sky models.

1. Introduction

Solar energy is abundant, renewable, widely distributed, safe, and clean, making it an ideal new energy source for power systems. In order to integrate solar energy into the power grid, an evaluation of the radiance input to solar power systems is required in many applications. Solar radiance may be attenuated by clouds and aerosols, thus the input of solar power system has some uncertainty. Clear-sky models describe the maximum input of solar power systems, which are particularly important for the design and sizing of solar power systems [1]. Besides, clear sky models can be used for solar irradiance forecasting and numerical weather prediction [2].

In the literature, simple clear sky models only use geometric variables to model the annual cycle of solar irradiance [3]. Daneshyar-Paltridge-Proctor model [4][5] uses zenith angle for calculation. Meinel model [6] includes zenith angle and extra-terrestrial solar irradiance. Laue model [7] adds elevation to modify Meinel model. Complex clear sky models take atmospheric variables into consideration to describe the solar irradiance attenuated by aerosols. Ineichen model [8] uses zenith angle, elevation, extra-terrestrial solar irradiance and Linke turbidity factor to calculate direct normal irradiance. Reference Evaluation of Solar Transmittance (REST) model [9] takes precipitable water, ozone amount, nitrogen dioxide amount, Angstrom turbidity coefficients and site pressure as inputs, and respectively calculates water vapor absorption, ozone absorption, nitrogen dioxide absorption, Rayleigh scattering, uniformly mixed gases absorption and aerosol extinction. REST2 model [1] separates the light wave into 2 bands, which gains more accuracy. There are many other complex clear sky models [10][11] such as Bird model [12][13], Simplified Solis model [14], Multilayer-weighted transmittance (MLWT) model [15][16].

The clear sky models mentioned above are empirical models or semi-empirical models. This article proposes a novel data-driven clear sky model for direct normal irradiance. In order to make use of the information in historical big data, a clear sky detection algorithm based on direct normal irradiance and air mass is designed in this paper. Firstly, clear-sky data is detected from historical big data automatically. After that, the weighted k-Nearest Neighbors regression algorithm is used to calculate clear-sky direct normal irradiance by historical clear-sky data.

The rest parts of this paper are organized as follows: in section 2, a clear sky detection algorithm is described in detail; a data-driven clear sky model is proposed based on weighted k-Nearest Neighbors regression algorithm using historical clear-sky data in section 3; the experiments and results using NERL data set are shown in section 4 and conclusions are drawn in section 5.

2. Clear sky detection

In order to take advantage of the useful information about clear sky in the historical big data, a clear sky detection algorithm is designed in this section. As shown in figure 1, the blue line represents direct normal irradiance (DNI) curve about time. Before 12:00 is clear sky, and after 13:00 is cloudy.

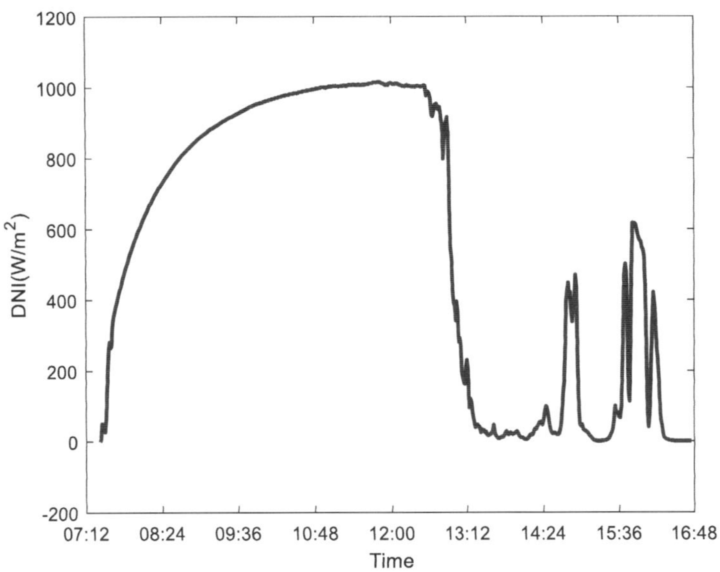

Figure 1. Direct normal irradiance curve.

By observing DNI curve in figure 1, clear sky data is detected by the following steps:

- Estimate aerosol extinction coefficient

  The Beer-Lambert law reveals the relationship between DNI and aerosol extinction coefficient at some wave length λ, as equation (1).

$$E(\lambda) = E_{toa}(\lambda) \cdot e^{-\tau_\lambda AM(\theta_z)} \tag{1}$$

  Ignoring the difference about aerosol extinction coefficient at any wave length, and take logarithm on both sides of the formula, the aerosol extinction coefficient can be approximately calculated in equation (2).

$$\ln E = \ln I_0 - \tau AM(\theta_z) \tag{2}$$

  where I_0 is the total solar irradiance, varying from 1363 W/m^2 to 1368 W/m^2. The I_0 can be approximately represented by the average of total solar irradiance. Therefore, the aerosol extinction coefficient τ can be calculated if DNI and airmass are given. It is obvious that the larger the aerosol extinction coefficient is, the more attenuation is caused by aerosols, and the less DNI reaches at the ground. Hence, for clear sky, the aerosol extinction coefficient is not larger than a certain value.

- Estimate convexity
  We can see obviously in figure 1 that before 12:00, clear sky DNI curve is approximately convex function about time, allowing a small range of noise; after 13:00, cloudy DNI curve fluctuates greatly and hardly has convexity. Therefore, we can estimate convexity to distinguish clear sky DNI curve from cloudy DNI curve.
- Estimate duration
  Clear sky duration is not less than a certain length of time. After two steps above, if the aerosol extinction coefficient is not larger than a certain value and the DNI curve satisfies convexity for some time, we also need to estimate whether this duration is less than a certain length of time, for example, 1 hour.
  After the steps above, the samples with label one can be considered as clear sky data.

3. A data-driven clear sky model

Considering the annual cycle of solar orbit and the earth's climate [17], the clear-sky direct normal irradiance has obvious seasonal difference. In this paper, we assume that the direct normal irradiance in the same month and day may have the similar pattern [18].

In this section, a data driven clear sky model is proposed based on the weighted k-Nearest Neighbors regression algorithm [19]. Firstly, historical clear sky direct normal irradiance data is detected and collected by the method described in section 2. Then the weighted k-Nearest Neighbors (k-NN) regression algorithm is used to calculate the clear-sky direct normal irradiance (DNI). The clear sky model is shown in figure 2.

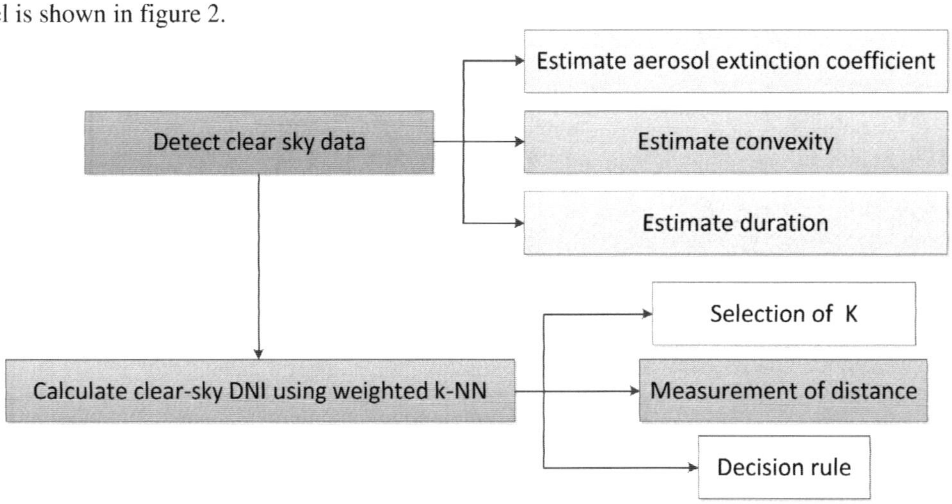

Figure 2. The data-driven clear sky model.

The selection of k, the distance measurement and the decision rule are three basic elements of the k-Nearest Neighbor algorithm.

- The selection of k
  In this paper, the value of k takes a relatively small number, and then gradually increases, using cross validation to select the optimal value of k.
- The distance measurement
  We define the distance from the sampling year to the current year as equation (3).

$$d_i = currentyear - samplingyear_i \tag{3}$$

  Where d_i is the distance of sample i, and $samplingyear_i$ is the year of sample i.
- The decision rule
  We define the decision rule as equation (4) and equation (5).

$$w_i = 1/d_i \tag{4}$$

$$DNI = \frac{\sum_{i=1}^{k} w_i \cdot y_i}{\sum_{i=1}^{k} w_i} \tag{5}$$

Where w_i is the weight of sample i, and y_i is the DNI value of sample i.

After calculating the clear-sky direct normal irradiance by weighted k-Nearest Neighbors regression algorithm, we can get a DNI curve for each day. The historical clear-sky data may be not continuous in some cases, thus the DNI calculated by historical data may not be satisfying. We modify the DNI by handling outliers and missing values, then regain convexity of the curve.

4. Experiments and results

4.1. Data collection

Data analyzed in this paper is provided by Solar Radiation Research Laboratory, which belong to National Renewable Energy Laboratory. The local information is shown in table 1.

Table 1: the location of Solar Radiation Research Laboratory.

Latitude	Longitude	Elevation	Time Zone
39.74°N	105.18°W	1829m	GMT-7

In this paper, data from 2004 to 2017 at 1-min sampling rate is used, including direct normal irradiance (DNI) and zenith angle.

4.2. Performance of clear sky detection

Define the threshold of aerosol extinction coefficient as 0.5, the threshold of noise as 10 W/m^2, and threshold of duration as 1 hour. Clear sky detection for 2017-01-01 is shown in figure 3, the dotted blue line denotes the original measured DNI. If it is clear-sky data, the solid red line coincide exactly with the blue line. Otherwise, the solid red line falls to zero.

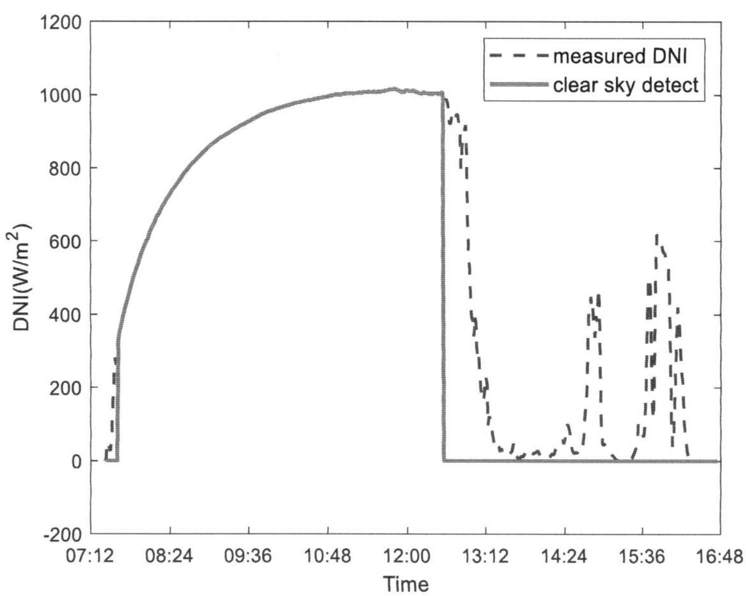

Figure 3. Clear sky detection.

4.3. Performance of the data-driven clear sky model

Experiments of the data-driven clear sky model proposed in this paper are conducted to evaluate the performance. For comparison, we adopt other three existing clear sky models, including a simple model (Meinel) and two complex models (Ineichen and REST).

The Meinel mode [6] is defined as equation (6).

$$E = I_0 \times 0.7^{AM \wedge 0.678} \tag{6}$$

The Ineichen model [8] is defined as equation (7).

$$E = b \cdot I_0 \cdot \exp(-0.09 \cdot AM \cdot (T_{LK} - 1)) \tag{7}$$

where T_{LK} is Linke turbidity factor, $b = 0.664 + 0.163 / f_h$ and $f_h = \exp(-altitude / 8000)$.

The REST model [9] is defined as equation (8).

$$E = T_R \cdot T_g \cdot T_o \cdot T_n \cdot T_w \cdot T_a \cdot I_0 \tag{8}$$

where T_R, T_g, T_o, T_n, T_w and T_a are the band transmittances for Rayleigh scattering, uniformly mixed gases absorption, ozone absorption, nitrogen dioxide absorption, water vapor absorption and aerosol extinction, respectively.

We use normalized root mean square error $nRMSE$, normalized mean absolute error $nMAE$ and normalized mean bias error $nMBE$ to evaluate the performance of each model.

Normalized root mean square error is defined as equation (9).

$$nRMSE = \frac{\sqrt{\frac{1}{N}\sum_{i=1}^{N}(E_{ci} - E_{mi})^2}}{\bar{E}_m} \times 100\% \tag{9}$$

Normalized mean absolute error is defined as equation (10).

$$nMAE = \frac{\frac{1}{N}\sum_{i=1}^{N}| E_{ci} - E_{mi} |}{\bar{E}_m} \times 100\% \tag{10}$$

Normalized mean bias error is defined as equation (11).

$$nMBE = \frac{\frac{1}{N}\sum_{i=1}^{N}(E_{ci} - E_{mi})}{\bar{E}_m} \times 100\% \tag{11}$$

In the formulas above, N is the number of samples, E_{ci} represents the calculated DNI by sample i, E_{mi} denotes the measured DNI and $\bar{E}_m$ represents the average of measured DNI.

Table 2. Performance evaluation of clear sky models.

Year	Models	nRMSE (%)	nMAE (%)	nMBE (%)
2015	Meinel	13.45	10.91	-5.37
	Ineichen	6.22	4.98	2.40
	REST	5.56	4.90	1.11
	Data-driven	**4.44**	**3.47**	**-0.84**
2016	Meinel	14.42	12.00	-2.44
	Ineichen	8.61	7.18	2.18
	REST	6.40	5.49	2.60
	Data-driven	**6.19**	**5.05**	**-1.21**
2017	Meinel	15.73	13.38	-3.61
	Ineichen	8.35	7.13	1.41
	REST	6.26	5.72	**0.95**
	Data-driven	**4.89**	**3.67**	-1.87

Select $k = 10$, and Table 2 lists the results of calculating clear sky DNI from 2015 to 2017, with three existing models and the data-driven model proposed in this paper. The best $nRMSE$, $nMAE$ and $nMBE$ for each year are shown in bold. Plots of measured DNI, the data-driven model predicted DNI and other three existing models predicted DNI for 2017-01-17 are shown in figure 4.

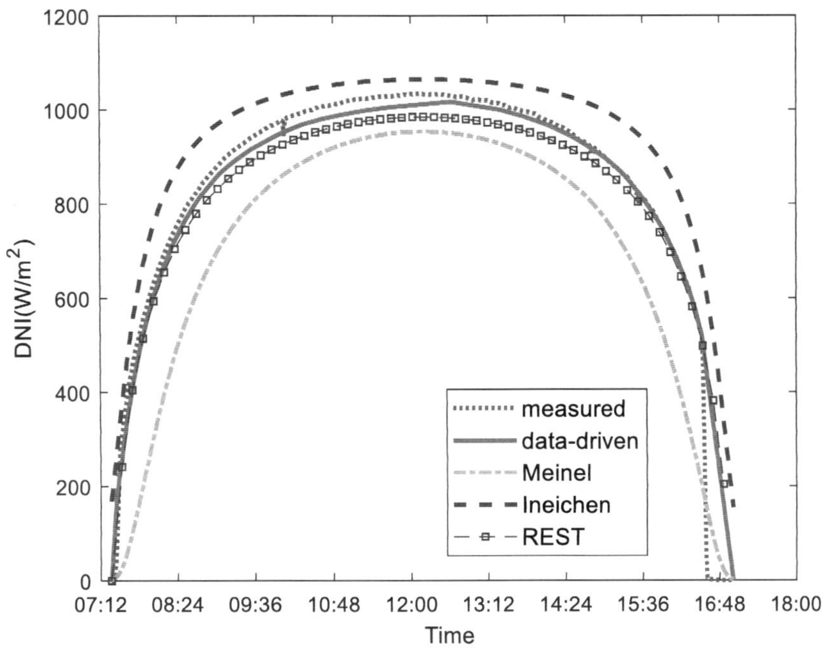

Figure 4. DNI forecasting plot for 2017-01-17

The results of the experiments show that the data-driven model proposed in this paper achieves the least $nRMSE$ and $nMAE$, performs better than other three clear sky models.

5. Conclusions

In this paper, a data-driven clear sky model is proposed for direct normal irradiance. Firstly, a clear sky detection algorithm is designed to collect clear sky data from historical big data automatically. Then, the weighted k-Nearest Neighbors regression algorithm is used to calculate the clear sky direct normal irradiance. Experiments are conducted using the database of NREL. The performance of the proposed model and three other clear sky models are compared, including Meinel, Ineichen and REST. The results of the experiments show that the data-driven model proposed in this paper achieves the least $nRMSE$ and $nMAE$, performs better than other three clear sky models.

6. References

[1] Gueymard C A 2008 *J. Solar Energy. REST2: High-performance solar radiation model for cloudless-sky irradiance, illuminance, and photosynthetically active radiation – Validation with a benchmark dataset.* **82(3)** 272-85

[2] Law E W, Prasad A A and Kay M 2014 *J. Solar Energy. Direct normal irradiance forecasting and its application to concentrated solar thermal output forecasting – A review.* **108** 287-307

[3] Stein J S, Hansen C W and Reno M J. 2012 *J. Sand. Global horizontal irradiance clear sky models: implementation and analysis.*

[4] Daneshyar M 1978 *J. Solar Energy. Solar radiation statistics for Iran.* **21** 345-49

[5] Paltridge G W and Proctor D 1976 *J. Solar Energy. Monthly mean solar radiation statistics for Australia.* **18** 235-43

[6] Meinel A B and Meinel M P 1976 *Applied solar energy* (MA: Addison-Wesley)
[7] Laue E G 1970 *J. Solar Energy. The measurement of solar spectral irradiance at different terrestrial elevations.* **13** 43-50
[8] Ineichen P and Perez R 2002 *J. Solar Energy. A new airmass independent formulation for the Linke turbidity coefficient.* **73(3)** 151-57
[9] Gueymard C A 2003 *J. Solar Energy. Direct solar transmittance and irradiance predictions with broadband models. Part I: detailed theoretical performance assessmen.* **74(5)** 355-79
[10] Inman R H, Edson J G and Coimbra C F M 2015 *J. Solar Energy. Impact of local broadband turbidity estimation on forecasting of clear sky direct normal irradiance.* **117** 125-38
[11] Janjai S, Sricharoen K and Pattarapanitchai S. 2011 *J. Applied Energy. Semi-empirical models for the estimation of clear sky solar global and direct normal irradiances in the tropics.* **88(12)** 4749-55
[12] Bird R E and Hulstrom R L 1980 *J. Solar Energy. Direct insolation models.*
[13] Bird R E and Hulstrom R L 1981 *J. Solar Energy. Simplified clear sky model for direct and diffuse insolation horizontal surfaces.*
[14] Ineichen P A 2008 *J. Solar Energy. broadband simplified version of the Solis clear sky model.* **82(8)** 758-62
[15] Gueymard C 1996 *Conf.Solar. Multilayer-weighted transmittance functions for use in broadband irradiance and turbidity calculations.* 281-8
[16] Gueymard C A 2010 *J. Applied Meteorology. Turbidity Determination from Broadband Irradiance Measurements: A Detailed Multicoefficient Approach.* 37(4) 414-35
[17] Zhu T T, Wei H K, Zhao X, Zhang C and Zhang K J 2017 *J. Renewable Energy. Clear sky model for wavelet forecast of direct normal irradiance.* **104** 1–8
[18] Zhang J X, Ehinger K A, Wei H K, Zhang K J and Yang J 2017 *J. Pattern Recognition. A novel graph-based optimization framework for salient object detection.* **64** 39–50
[19] Harrington P 2012 *Machine Learning in Action* (Manning)

Acknowledgments
The authors gratefully acknowledge the National Renewable Energy Laboratory providing all data in this research. This work is supported by the National Natural Science Foundation of China (Grant No.61374006, Grant No.61773118 and Grant No.61703100), and by the Fundamental Research Funds for the Central Universities.

Predicting residential electricity consumption using neural networks: *A case study*

Sang Guun Yoo[1,2,3,*] and Myriam Hernández-Álvarez[1]

[1] Departamento de Informática y Ciencias de la Computación, Escuela Politécnica Nacional, Ladrón de Guevara, E11-253, Quito, Ecuador
[2] Smart Lab, Escuela Politécnica Nacional, Ladrón de Guevara, E11-253, Quito, Ecuador
[3] Departamento de Ciencias de la Computación, Universidad de las Fuerzas Armadas ESPE, Av. General Rumiñahui s/n, Sangolquí, Ecuador

Email: *sang.yoo@epn.edu.ec

Abstract. Electricity demand prediction plays an essential role in short-term load allocation and long-term planning for new generation and transmission infrastructures. An accurate prediction also allows to take better decisions in terms of cost and energy efficiency. In this aspect, this paper proposes a model for predicting the electricity consumption of residential area in Seoul using neural network. This work has analyzed several particular characteristics of the aforementioned city to extract variables that could have direct influence in the electricity consumption pattern. Using the extracted variables, this paper could forecast the residential electricity consumption with an average error rate of 2.0375% and it could demonstrate how the elderly population is the parameter that influences with major weight at the moment of forecasting residential electricity consumption. Additionally, the presented work illustrates how executing the supervised learning process using a data set organized by months of the year can reduce considerably the error rate. Furthermore, the analysis or results delivers interesting findings related to the energy consumption in Seoul.

1. Introduction

Electricity demand prediction plays an essential role in short-term load allocation and long-term planning for new generation and transmission infrastructures. An accurate prediction also allows to take better decisions in terms of cost and energy efficiency. In this aspect, responding to the significance of forecasting electricity demand, several works have dedicated their content to this field [1-7]. Some of those works have predicted the energy consumption based on weather variables (e.g. temperature and relative humidity, or season of the year) [3, 4] and others forecasted based on socio-economic and demographic variables [6, 7].

Even though it is important to predict the electricity consumption based on general variables, we believe that it is possible to have a better result if we understand the particular characteristics of the population and zone where the electricity is consumed. To confirm this assumption, this paper proposes the prediction of residential electricity consumption of a specific city i.e. Seoul, the capital city of South Korea, based on real data and unique characteristics of the aforementioned place.

Seoul (officially the Seoul Special City) is the capital and largest city of Republic of Korea. Seoul is considered one of the largest cities in the world with approximately 10 million people living in the city, 12.7 million people living in the urban area, and 25.5 million people living in the metropolitan area.

Additionally, Seoul is considered as a leading and growing global city, resulting from an economic and financial explosion called the Miracle on the Han River which converted it into the fourth biggest metropolitan economy of the world with a GDP of US$845.9 billion in 2014 just behind Tokyo, New York City and Los Angeles.
Even though Seoul has many similarities with most of megacities in the world, it has several particular characteristics (see section 2 for more details). The authors of this work believes that those particular characteristics have a direct effect in the residential electricity consumption in Seoul. They also believe that such particularities can explain the trend change of residential electricity consumption occurred in the last few years (consumption decrease after decades of constant growing) (see Figure 1).

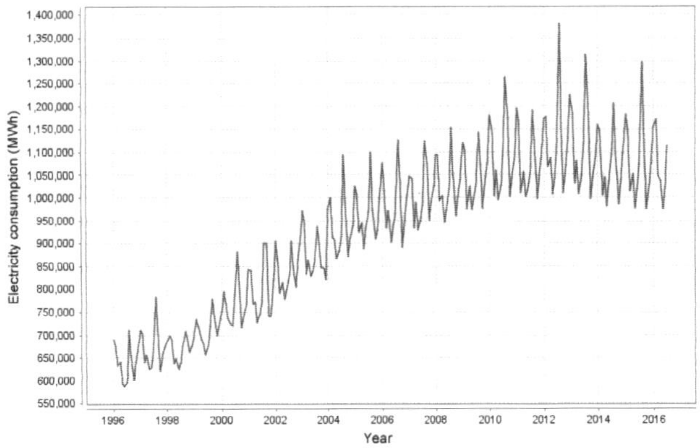

Figure 1. Residential electricity consumption in Seoul, Republic of Korea

With this antecedent, this work proposes the prediction of the residential electricity consumption using variables extracted from aforementioned characteristics. This paper contributes in the field of electricity consumption forecasting in several ways. (1) First, this paper delivers a neural network prediction model for Seoul based on real data. (2) It also analyses the performance of the model when using different data sets (training data grouped by months of the year or single block). (3) Finally, this work makes an analysis of different variables that affect the prediction and determines which of them have the highest level of influence.
The rest of the paper is organized as follows. Section 2 and 3 deliver the hypothesis to be proved in this work and the way how they are going to be proved (methods and metrics). Then, Section 4 analyses the results obtained after executing the modelling process. Finally, Section 5 concludes this paper.

2. Research Scheme and Hypothesis

The research scheme to be used is presented in figure 2. As mentioned before, this work analyses the particular characteristics of Seoul and uses them as variables to predict its residential electricity consumption. Again, those particular characteristics are as follows.

(1) First, Koreans (including people living in Seoul) are working 1.2 times longer on average than their peers in other developed countries. According to the Organization for Economic Cooperation and Development (OECD), Koreans worked on average 2113 hours in 2015; 347 hours more than the OECD average, 323 hours longer than people from United States, and 742 hours longer than Germans [8].

(2) According to the Ministry of Health, Welfare and Family Affairs of Korea, Korean young students study an average of three hours more per day than teenagers in other OECD countries [9]. The Korean adolescents also spent on private tutoring on average one hour and 59 minutes per week which is leaded by a large margin the 22 minutes in Japan, 19 minutes in the United States, 16 minutes in the United Kingdom, and three minutes in Finland [9].

(3) Since several years ago, South Korea is experiencing extremely low fertility and a rapidly increasing number of elderly people. The inequitable social and economic situation in Korean have encouraged people to delay marriage and childbearing [10]. Statistics also indicates that single-person households are increasing rapidly [11].

(4) Seoul is a city with four distinct seasons, and therefore, its weather varies depending on the period of the year. In summer, there is a hot and humid weather with temperatures elevated as high as 35 °C because of the influence of the North Pacific high-pressure system. In winter, the temperature and humidity (as low as -20 °C) drops influenced by Siberian high-pressure and west winds.

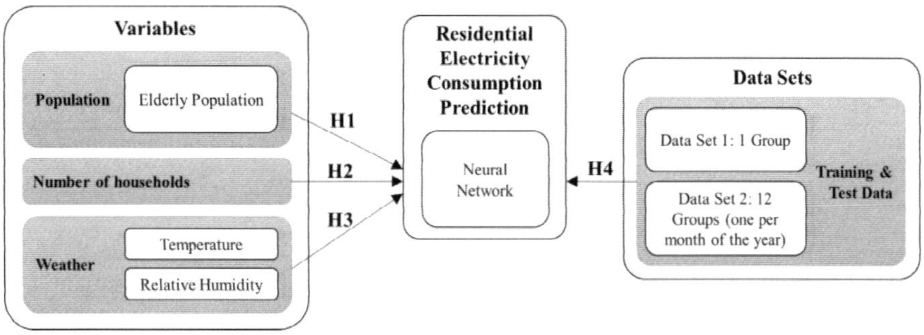

Figure 2. The Proposed Research Scheme

Based on the aforementioned data, it is possible to deduce that students (young population) spent most of their time in schools and private tutoring, while the economically active adults spend most of their time at work. Consequently, it is possible to reason that that elder people are the segment of the population that spends more time at their home. Having this background, this paper hypothesizes the following: H1: Elderly population has an important influence when predicting the electricity consumptions in residential areas of Seoul. Additionally, since the single-person households are increasing, it is possible to hypothesize the following: H2: Number of households influences when predicting the electricity consumptions in residential areas in Seoul. Finally, since Seoul has four different seasons with high variation of temperature and humidity, electricity consumption pattern will depend on the period of the year. Therefore, this work also hypothesizes: H3: Temperature and relative humidity influence when predicting the electricity consumptions in residential areas in Seoul, H4: Electricity prediction based on months of the year will be more precise than the yearly based prediction. In summary, the intention of this work is to confirm the influence of several factors i.e. total population, elderly population, weather (temperature and humidity) and number of households in predicting the electrical consumption in residential areas of Seoul, and to understand which of those factors are the most influential. Additionally, this work pretends to show that prediction based on months of the year is a better option than a prediction based on a yearly organized data set.

3. Research Method

3.1. Data

For forecasting the residential electricity consumption in Seoul, we have collected real data of approximately 20 years i.e. from January 1996 to July 2016. Seoul´s demographic and residential electricity consumption data were delivered by the Seoul Open Data Plaza [12], while the weather data were provided by the Korea Meteorological Administration [13]. All the accumulated data was normalized to use only the following attributes (to be used as variables): date, total population (called as population in the rest of the paper), elderly population (defined by OECD as people aged 65 and over [14]), average temperature (called as temperature in the rest of the paper), average relative humidity (called as humidity in the rest of the paper), and number of households (called as households in the rest of the paper). Once normalized the collected data, it was used to create two different data sets. (1) The first data set was created by dividing the filtered data into training data (data from January 1996 to December 2012) and test data (data from January 2013 to July 2016) (see Figure 3). (2) On the other hand, the second data set was created by dividing the normalized data per month of the year (see Figure 4) and separating them into two groups: one for training purposes (data from January 1996 to December 2012) and another for performance testing purposes (data from January 2013 to July 2016) (see figure 4).

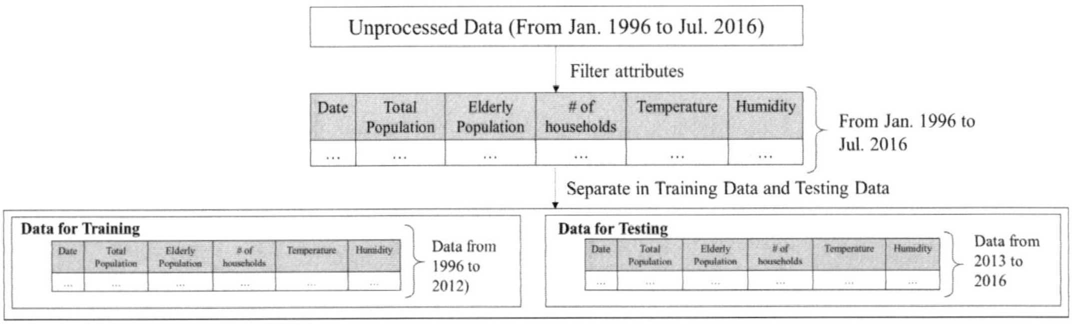

Figure 3. Data Set 1

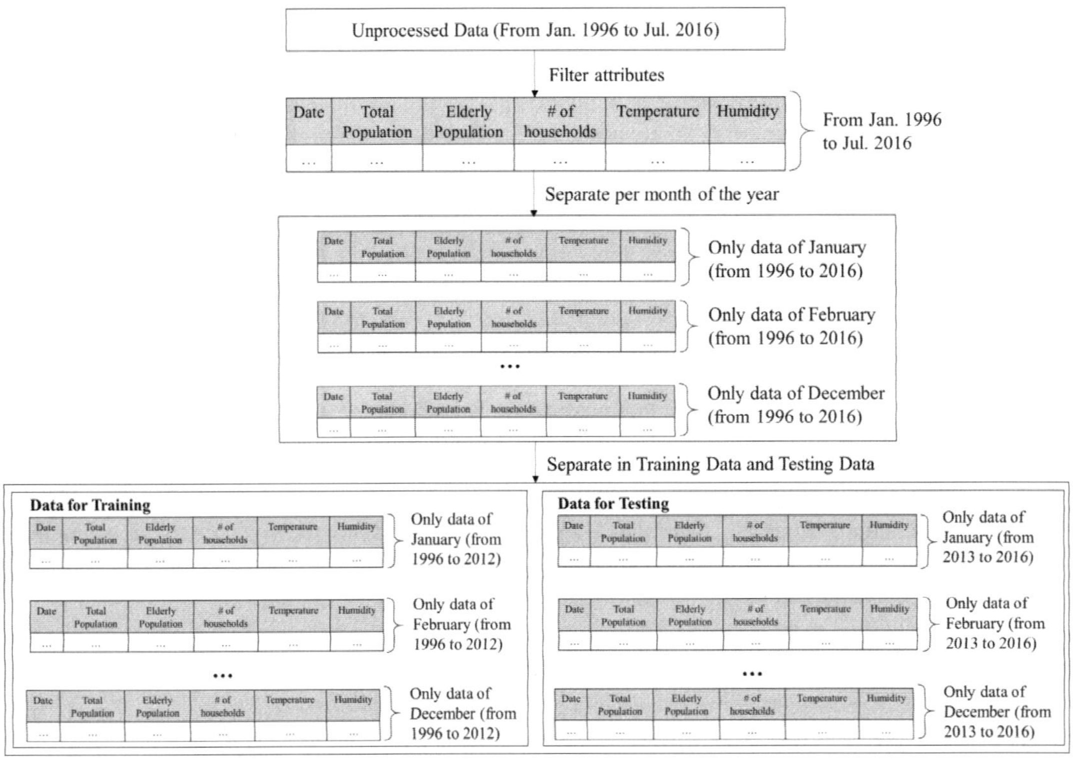

Figure 4. Data Set 2: Data separated per months of the year

3.2. Machine Learning Model and Variables

For this work, we have applied the neural network, since it was the best supervised machine learning model in predicting the electrical energy consumption in Seoul. We executed the prediction process using other models, such as linear regression, polynomial regression, and support vector machine, but the obtained results were less exact than the results gotten using neural networks. Furthermore, the reasons behind selecting the neural network are its capability to predict future values of more than one variable at the same time and to model the nonlinear relation in the data structure, and because their widespread usage for predicting values [15-17].

Once selected the machine learning mechanism, we defined the variables to use. As mentioned before, this work has selected five variables i.e. temperature, humidity, households, population, and elderly population. To understand, which of those variables has more influence in forecasting the electricity consumption, we have executed the prediction model using all possible combinations i.e. (1) using only elderly population, (2) using only population, (3) using population and elderly population, (4) using only households, (5) using households and elderly population, (6) using population and households, …, (31) using temperature, humidity, households, population and elderly population.

3.3. Performance Metrics

This work has used the average of three metrics i.e. coefficient of variance, mean bias error, and mean absolute percentage of error to measure the performance of the prediction model. This work has decided to use the average of the aforementioned metrics to reduce the limitations that each one of those metrics could have. The details of such metrics are described below.

1) Coefficient of Variance (CV): It defines how much the prediction error varies with respect to the target's mean [18]. Lower value means lower error rate. It is calculated as follows:

$$CV = \frac{\left(\frac{1}{N-1} \sum_{i=1}^{N} (y_i - \hat{y}_i)^2 \right)^{1/2}}{\overline{y}} \times 100 \tag{1}$$

where $\hat{y}_i$ is the predicted residential electricity consumption, y_i is the real residential electricity consumption, and is the average residential electricity consumption.

2) Mean Bias Error (MBE): It indicates how a particular model is to over-estimate or under-estimate the actual electricity consumption. Lower value means lower error rate. MBE is calculated as follows:

$$MBE = \frac{\frac{1}{N-1} \sum_{i=1}^{N} (y_i - \hat{y}_i)}{\overline{y}} \times 100 \tag{2}$$

3) Mean Absolute Percentage of Error (MAPE): It indicates the percentage of error per prediction [3, 19]. MAPE is calculated as follows:

$$MAPE = \frac{1}{N} \sum_{i=1}^{N} \frac{|y_i - \hat{y}_i|}{y_i}) \times 100 \tag{3}$$

4. Data Analysis and Results

4.1. Analysis of Data Set 1

Table 1 illustrates the prediction results of the neural network of all possible variable combinations using data set 1. Such table shows the combinations of variables and the resulted performance metrics (CV, MBE, MAPE, and average). It also indicates which variable combinations had the lowest rate of error i.e. "order". Since not all the variable combinations resulted with the best error rate, we have selected the three best variable combinations (10% of the set) for analysis (see Table 2). This analysis shows how the lowest average error rate (4.26%) is reached applying the variable combination {temperature, population, elderly population}. The comparison of real consumption against the prediction value is detailed in figure 5.

4.2. Analysis of Data Set 2

Table 3 illustrates the neural network's prediction results of all possible variable combinations using data set 2. Such table shows the combinations of variables and the resulted performance metrics (CV, MBE, MAPE, and average). It also indicates which variable combinations had the lowest rate of error i.e. "order". Since not all the combination resulted with the best error rate, we have highlighted the three best variable combinations (10% of the set) of each month of the year for analysis (see Table 3). The analysis of Table 3 allowed the construction of Table 4 with the variables combinations having the three lowest average error rates for each month of the year. The comparison of real electricity consumption value against the prediction value of the best variable combination of each month of the year is detailed in figure 6.

4.3. Comparison of Prediction Executed using Data Set 1 and Data Set 2

Comparing Table 2 and Table 4, and comparing Figure 5 and Figure 6, it is possible to deduce which data set is better for prediction. The results indicate that the error rates of Table 4 are much lower than error rates of Table 2, which indicates that usage of data separated by months of the year has superior performance. The superiority of prediction using Data Set 2 is also visualized in Figure 6, which shows how the prediction line is much more close to the real values than in Figure 5. This analysis corroborates the hypothesis H4 which indicates "Electricity prediction based on months of the year will be more precise than the yearly based prediction".

4.4. Analysis of Variable Combinations

After verifying the superiority of prediction using data set 2 (per months of the year), we have analyzed which variables were the most influential in forecasting the electricity consumption in residential areas. For this, we have selected the 20% of combinations (6 combinations) that delivered the lowest average error rate and verified which variables were more frequently used in electricity consumption predictions. As shown in Table 5, this analysis indicates that the most influential variables for January are elderly population and population (5 repetitions each), for February are population (5 repetitions) and elderly population/temperature/households (3 repetitions each), for March are humidity/temperature (4 repetitions each) and elderly population (3 repetitions), for April are humidity/temperature (4 repetitions each) and elderly population/humidity (3 repetitions each), for May are elderly population and temperature (4 repetitions each), for June are elderly population (4 repetitions) and humidity (3 repetitions), for July are temperature (6 repetitions) and elderly population/households (4 repetitions each), for August are temperature and elderly population (3 repetitions each), for September are elderly population (5 repetitions) and populations/humidity/temperature (4 repetitions each), for October are number of households (5 repetitions) and elderly population/population (3 repetitions each), for November are population (5 repetitions) and elderly population (4 repetitions), and for December are temperature (4 repetitions), humidity (3 repetitions) and elderly populations (2 repetitions).

This data indicates that depending on the month of the year all the variables have a direct influence in predicting the energy consumption in residential areas of Seoul. This situation corroborates the hypothesis H2: Number of households has influence predicting the electricity consumptions in residential areas in Seoul and H3: Temperature and relative humidity has influence predicting the electricity consumptions in residential areas in Seoul.

Table 5 also shows how the most influential variables for the whole year are the elderly population (43 repetitions), temperature (40 repetitions), and population (30 repetitions). This show how the elderly population is the variable that has the biggest influence in electricity consumption in residential areas of Seoul. Therefore, it is important to take account this parameter for the electricity consumption planning in the metropolitan area of Seoul. This corroborates the correctness of hypothesis H1 which says "Elderly population has important influence in predicting the electricity consumptions in residential areas of Seoul".

Figure 7 corroborates the hypothesis H1. The total population decrease does not affect directly to the residential electricity consumption since population decrease is neutralized by the elderly population increase.

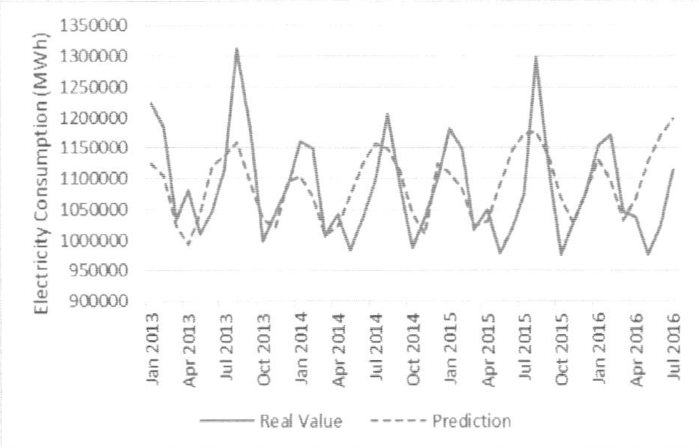

Figure 5. Real Consumption vs Prediction in Best Variables Combination using Data Set 1

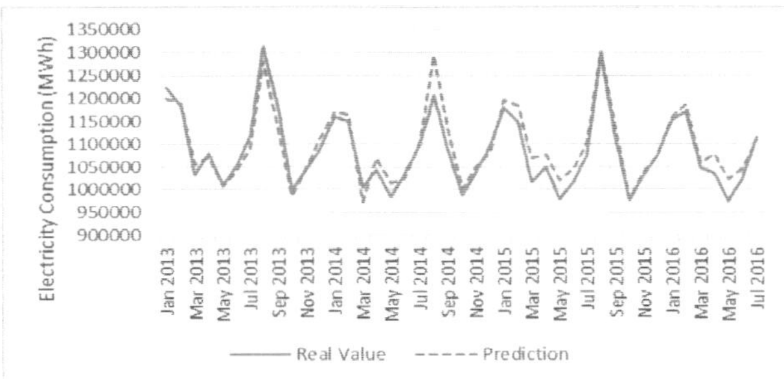

Figure 6. Real Consumption vs Prediction in Best Variables Combination using Data Set 2

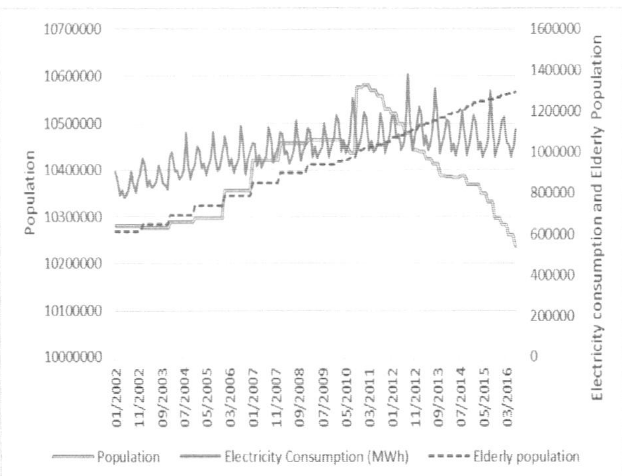

Figure 7. Population vs. Elderly Population in Residential Electricity Consumption

5. Conclusions

This paper has proposed a model for predicting the electricity consumption of residential area in Seoul. This work has analysed several variables that could have direct influence in electricity consumption based on the particularities of the mentioned city. The presented work has demonstrate how the elderly population is the variable that has the biggest influence at the moment of forecasting residential electricity consumption. Additionally, it has shown how the analysis of data separated by months of the year can reduce considerably the error rate. Furthermore, this work has delivered interesting findings related to the energy consumption forecast in Seoul.

6. References

[1] Hamzacebi C 2007 Forecastirng of Turkey's net electricity energy consumption on sectoral bases *Energy Policy* vol 35 pp 2009-2016
[2] Limanond T. et al. 2011 Projection of future transport energy demand of Thailand *Energy Policy* vol 39 pp 2754-2763.
[3] Dong B et al. 2005 Applying support vector machine to predict building energy consumption in tropical region *Energy and Buildings* vol.37 pp 545-553
[4] Tso G and Yau K 2007 Predicting electricity energy consumption: A comparison of regression analysis, decision tree and neural networks *Energy* vol 20 pp 1761-1768
[5] Ekonomou L 2010 Greek long-term energy consumption prediction using artificial neural networks *Energy* vol 35 pp 512-517
[6] Kankal M et al. 2011 Modeling and forecasting of Turkey's energy consumption using socio-economic and demographic variables *Applied Energy* vol 88 pp 1927-1939
[7] Sözen A and Arcaklioglu E 2007 Prediction of net energy consumption based on economic indicators (GNP and GDP) in Turkey *Energy Policy* vol 25 pp 4981-4992
[8] OECD 2016 OECD Data [Online] Available: https://data.oecd.org/emp/hours-worked.htm [Accessed: November 23, 2016]
[9] Chosunilbo 2009 Korean Youth Study Longest Hours in OECD. [Online] Available: http://english.chosun.com/site/data/html_dir/2009/08/10/2009081000200.html [Accessed: November 23, 2016]
[10] Hervery E 2012 Bracing for Low Fertility and a Large Elderly Population in South Korea *Korea Economic Institute Academic Paper Series* April 18.
[11] The Korea Times 2016 Korea faces rapidly aging population [Online] Available: http://www.koreatimes.co.kr/www/news/biz/2016/03/123_201016.html [Accessed: November 23, 2016]
[12] Seoul Metropolitan Government 2016 Seoul Open Data Plaza [Online] Available: http://data.seoul.go.kr [Accessed: November 23, 2016]
[13] Korea Metrological Administration 2016 [Online] Available: http://www.kma.go.kr [Accessed: November 23, 2016]
[14] OECD 2014 Elderly population [Online] Available: https://data.oecd.org/pop/elderly-population.htm [Accessed: November 23, 2016]
[15] Geem Z 2011 Transport energy demand modeling of South Korea using artificial neural network *Energy Policy* vol 39 pp 4644-4650
[16] Lisboa P and Taktak A 2006 The use of artificial neural networks in decision support in cancer: a systematic review *Neural Networks* vol 19 pp 408–415
[17] Murat Y and Ceylan H 2006 Use of artificial neural networks for transport energy demand modeling *Energy Policy* vol 34(17) pp 3165-3172.
[18] Kreider J and Haberl J 1994 Predicting hourly building energy use: the great energy predictor shootout – overview and discussion of results *ASHRAE Transactions* vol 100 (2) pp 1104–1118
[19] Gonzalez P and Zamarreno J 2005 Prediction of hourly energy consumption in buildings based on a feedback artifitial neural network *Energy and Buildings* vol 37 (6) pp 595-601

Acknowledgments

The authors gratefully acknowledge the financial support provided by the Escuela Politécnica Nacional, for the development of the project PIJ-17-08 - "Diseño e implementación de un sistema de parqueadero inteligente".

Table 1. Combinations of Variables and Corresponding Metrics Results using Data Sample 1 (NE=Not Executed)

Variables		Variables combinations															
		1	**2**	**3**	**4**	**5**	**6**	**7**	**8**	**9**	**10**	**11**	**12**	**13**	**14**	**15**	**16**
	Humidity	no	no	no	no	no	no	no	no	no	no	no	no	no	no	no	no
	Temperature	no	no	no	no	no	no	no	no	yes	yes	yes	yes	yes	yes	yes	yes
	# of households	no	no	no	no	yes	yes	yes	yes	no	no	no	no	yes	yes	yes	yes
	Population	no	no	yes	yes	no	no	yes	yes	no	no	yes	yes	no	no	yes	yes
	Elderly pop.	no	yes	no	yes	no	yes	no	yes	no	yes	no	yes	no	yes	no	yes
Metric (In %)	**CV**	NE	17.95	10.16	7.74	9.75	8.69	7.80	13.06	6.95	7.82	7.81	6.85	7.85	8.18	8.79	14.23
	MBE	NE	15.61	6.69	0.64	5.83	2.78	0.33	8.68	4.21	0.34	0.50	0.46	0.15	2.71	1.07	11.68
	MAPE	NE	16.51	8.99	6.28	8.59	7.35	6.24	11.41	5.94	6.14	6.28	5.46	6.19	6.72	7.34	12.59
	Average	NE	16.69	8.62	4.88	8.06	6.27	4.79	11.05	5.70	4.77	4.86	4.26	4.73	5.87	5.74	12.83
	Order	NE	**27**	**21**	**13**	**20**	**19**	**8**	**24**	**16**	**7**	**11**	**1**	**5**	**18**	**17**	**25**

Variables		Variables combinations															
		17	**18**	**19**	**20**	**21**	**22**	**23**	**24**	**25**	**26**	**27**	**28**	**29**	**30**	**31**	**32**
	Humidity	yes	yes	yes	yes	yes	yes	yes	yes	yes	yes	yes	yes	yes	yes	yes	yes
	Temperature	no	no	no	no	no	no	no	no	yes	yes	yes	yes	yes	yes	yes	yes
	# of households	no	no	no	no	yes	yes	yes	yes	no	no	no	no	yes	yes	yes	yes
	Population	no	no	yes	yes	no	no	yes	yes	no	no	yes	yes	no	no	yes	yes
	Elderly pop.	no	yes	no	yes	no	yes	no	yes	no	yes	no	yes	no	yes	no	yes
Metric (In %)	**CV**	12.00	7.78	7.82	8.14	8.00	12.18	15.56	25.80	7.70	6.31	7.31	26.17	7.71	20.14	19.77	7.87
	MBE	8.51	0.50	0.49	1.19	1.65	6.51	10.03	23.26	0.03	1.87	2.05	22.01	0.86	18.60	17.91	0.12
	MAPE	10.45	6.26	6.28	6.25	6.68	10.27	13.12	23.58	6.04	4.85	5.18	22.17	5.49	18.95	18.29	6.23
	Average	10.32	4.85	4.87	5.19	5.44	9.65	12.90	24.21	4.59	4.34	4.85	23.45	4.69	19.23	18.66	4.74
	Order	**23**	**10**	**12**	**14**	**15**	**22**	**26**	**31**	**3**	**2**	**9**	**30**	**4**	**29**	**28**	**6**

Table 2. Three Combinations of Variables with Lowest Error Rates and Corresponding Metrics Results using Data Sample 1

Variables		Variables combinations		
		12	**26**	**25**
	Humidity	no	yes	yes
	Temperature	yes	yes	yes
	# of households	no	no	no
	Population	yes	no	no
	Elderly pop.	yes	yes	no
Metric (In %)	**CV**	6.85	6.31	7.70
	MBE	0.46	1.87	0.03
	MAPE	5.46	4.85	6.04
	Average	4.26	4.34	4.59
	Order	1	2	3

IOP Conf. Series: Journal of Physics: Conf. Series **1072** (2018) 012005 doi:10.1088/1742-6596/1072/1/012005

Table 3. Combinations of Variables and Corresponding Metrics Results using Data Sample 2

Variables		Variables combinations															
		1	2	3	4	5	6	7	8	9	10	11	12	13	14	15	16
	Humidity	no	no	no	no	no	no	no	no	no	no	no	no	no	no	no	no
	Temperature	no	no	no	no	no	no	no	no	yes	yes	yes	yes	yes	yes	yes	yes
	# of households	no	no	no	no	yes	yes	yes	yes	no	no	no	no	yes	yes	yes	yes
	Population	no	no	yes	yes	no	no	yes	yes	no	no	yes	yes	no	no	yes	yes
	Elderly pop.	no	yes	no	yes	no	yes	no	yes	no	yes	no	yes	no	yes	no	yes
Jan.	CV	NE	6.10	4.17	2.53	6.25	7.58	2.08	6.40	4.84	3.23	3.21	2.54	3.43	4.32	5.41	4.77
	MBE	NE	5.33	4.41	2.11	5.44	7.16	0.37	7.47	4.59	0.99	3.14	1.24	1.67	3.08	5.96	5.11
	MAPE	NE	4.85	3.28	1.72	4.96	5.78	1.32	5.59	3.69	2.62	2.33	2.01	2.89	3.67	4.47	3.80
	Average	NE	5.43	3.95	2.12	5.55	6.84	1.26	6.49	4.37	2.28	2.90	1.93	2.67	3.69	5.28	4.56
	Order	NE	23	19	5	24	26	3	25	20	6	13	4	11	18	22	21
Feb.	CV	NE	6.31	3.49	9.53	7.93	11.13	8.94	2.90	3.59	4.43	6.50	7.90	5.00	7.06	3.52	11.82
	MBE	NE	6.69	3.46	10.48	8.54	12.16	9.82	2.73	3.70	4.85	7.19	8.14	5.39	7.20	1.78	13.26
	MAPE	NE	5.04	2.61	7.88	6.43	9.15	7.39	2.03	2.79	3.65	5.41	6.14	4.06	5.43	2.62	9.97
	Average	NE	6.02	3.19	9.29	7.64	10.81	8.72	2.55	3.36	4.31	6.37	7.39	4.82	6.57	2.64	11.68
	Order	NE	14	5	28	21	29	26	2	6	8	15	19	11	16	3	31
Mar.	CV	NE	6.15	12.08	12.44	6.45	10.71	12.80	10.06	8.10	5.42	9.91	7.48	9.31	8.24	6.59	14.52
	MBE	NE	6.81	13.66	14.03	7.18	11.96	14.41	11.33	9.00	6.03	11.12	8.40	10.42	9.27	7.34	16.25
	MAPE	NE	5.13	10.26	10.54	5.40	8.98	10.82	8.51	6.77	4.54	8.35	6.31	7.83	6.97	5.52	12.20
	Average	NE	6.03	12.00	12.34	6.34	10.55	12.68	9.97	7.96	5.33	9.79	7.40	9.19	8.16	6.48	14.32
	Order	NE	5	24	25	8	22	26	21	15	3	20	12	19	16	9	28
Apr.	CV	NE	5.89	10.45	11.34	6.69	9.30	12.01	6.30	3.02	6.19	12.82	7.51	7.96	6.96	9.02	12.34
	MBE	NE	5.97	11.31	12.50	6.88	9.88	13.11	6.70	2.59	6.66	13.85	8.21	8.60	7.55	9.64	13.50
	MAPE	NE	4.52	8.53	9.42	5.20	7.46	9.89	5.06	2.41	5.03	10.45	6.19	6.48	5.70	7.28	10.18
	Average	NE	5.46	10.10	11.09	6.26	8.88	11.67	6.02	2.67	5.96	12.37	7.30	7.68	6.74	8.65	12.01
	Order	NE	5	24	27	9	21	29	7	1	6	31	15	17	11	18	30
May	CV	NE	7.17	9.90	5.66	8.09	5.03	6.47	4.87	7.39	4.34	5.65	4.50	5.63	5.05	4.63	8.11
	MBE	NE	7.71	10.87	5.85	8.72	4.97	6.66	4.87	7.98	4.26	5.86	4.40	5.71	5.09	4.50	8.77
	MAPE	NE	5.82	8.19	4.42	6.58	3.76	5.03	3.71	6.01	3.25	4.43	3.37	4.31	3.85	3.46	6.61
	Average	NE	6.90	9.65	5.31	7.80	4.58	6.05	4.49	7.13	3.95	5.31	4.09	5.22	4.66	4.20	7.83
	Order	NE	23	30	14	26	7	19	5	24	2	15	3	11	8	4	27
Jun.	CV	NE	3.69	9.59	7.97	5.38	5.93	11.44	9.19	2.93	3.54	11.82	3.56	5.54	3.63	5.02	3.72
	MBE	NE	3.38	10.37	8.81	5.37	6.09	12.56	10.01	2.96	3.68	13.03	3.64	5.86	3.66	5.23	3.84
	MAPE	NE	2.71	7.81	6.63	4.06	4.59	9.45	7.54	2.23	2.77	9.81	2.75	4.42	2.76	3.95	2.90
	Average	NE	3.26	9.26	7.80	4.94	5.54	11.15	8.92	2.71	3.33	11.55	3.32	5.27	3.35	4.73	3.49
	Order	NE	5	27	24	16	21	29	26	3	7	30	6	18	8	14	9
Jul.	CV	NE	4.02	9.76	12.38	6.01	6.01	13.91	12.45	7.51	3.46	8.32	5.68	4.08	1.97	3.08	2.46
	MBE	NE	3.77	10.52	13.61	6.14	6.14	15.13	13.78	7.72	2.85	8.72	5.99	3.57	0.06	2.50	1.38
	MAPE	NE	2.88	7.92	10.24	4.64	4.64	11.38	10.37	5.82	2.78	6.57	4.52	3.23	1.44	2.29	1.89
	Average	NE	3.56	9.40	12.08	5.60	5.60	13.47	12.20	7.02	3.03	7.87	5.39	3.63	1.16	2.62	1.91
	Order	NE	10	27	29	18	19	31	30	22	6	26	17	12	1	5	3
Aug.	CV	NE	5.48	22.84	36.45	5.20	11.68	35.02	26.73	5.92	8.16	20.53	21.60	13.58	18.11	6.58	37.64
	MBE	NE	3.28	26.83	43.95	2.46	12.89	41.89	32.93	2.98	0.88	25.09	26.04	14.97	22.00	7.81	44.05
	MAPE	NE	3.51	18.07	29.51	3.38	8.75	28.13	22.15	3.77	5.84	16.79	17.44	9.83	14.74	5.17	29.62
	Average	NE	4.09	22.58	36.64	3.68	11.11	35.01	27.27	4.22	4.96	20.80	21.69	12.79	18.28	6.52	37.10
	Order	NE	2	20	28	1	10	24	22	3	4	18	19	11	16	6	29
Sep.	CV	NE	7.45	6.97	6.66	7.47	7.88	7.36	11.72	6.88	7.93	9.35	4.52	7.86	5.36	7.20	19.63
	MBE	NE	6.80	6.22	6.04	6.82	7.61	7.19	13.03	5.92	7.67	9.61	1.02	7.50	3.17	6.34	22.58
	MAPE	NE	5.38	5.06	4.67	5.41	5.48	4.92	8.85	5.12	5.51	6.55	3.54	5.52	4.27	5.32	15.25
	Average	NE	6.55	6.09	5.79	6.57	6.99	6.49	11.20	5.97	7.04	8.51	3.03	6.96	4.27	6.29	19.15
	Order	NE	20	15	13	21	23	19	30	14	24	27	1	22	2	16	31
Oct.	CV	NE	5.41	6.28	3.25	5.46	2.51	2.45	1.19	3.88	2.12	2.95	3.91	2.77	5.05	2.91	5.53
	MBE	NE	6.27	7.42	3.62	6.31	2.31	2.61	0.80	4.36	1.99	3.14	4.50	2.84	5.87	3.12	6.43
	MAPE	NE	4.19	4.96	2.43	4.22	1.64	1.75	0.93	2.92	1.36	2.10	3.01	1.91	3.92	2.09	4.30
	Average	NE	5.29	6.22	3.10	5.33	2.15	2.27	0.97	3.72	1.82	2.73	3.81	2.51	4.95	2.70	5.42
	Order	NE	22	28	10	23	4	5	2	13	3	8	14	6	21	7	24
Nov.	CV	NE	6.34	6.96	0.88	6.66	5.10	1.30	7.36	4.77	4.71	4.49	4.11	5.01	5.34	4.67	6.03
	MBE	NE	7.55	8.38	1.02	7.91	5.96	1.47	8.88	5.67	5.52	5.33	4.91	5.86	6.32	5.48	7.20
	MAPE	NE	5.04	5.59	0.68	5.28	3.98	0.98	5.93	3.79	3.69	3.56	3.28	3.92	4.22	3.66	4.81
	Average	NE	6.31	6.98	0.86	6.62	5.02	1.25	7.39	4.74	4.64	4.46	4.10	4.93	5.29	4.61	6.01
	Order	NE	23	27	1	25	13	2	29	11	10	8	4	12	16	9	20
Dec.	CV	NE	4.02	7.49	2.66	4.02	8.49	4.21	15.04	2.09	3.05	3.29	6.49	2.58	3.66	4.48	4.44
	MBE	NE	4.56	8.80	3.15	4.55	10.11	4.86	18.22	0.56	3.10	2.56	7.79	2.85	2.50	4.04	5.22
	MAPE	NE	3.05	5.88	2.10	3.04	6.75	3.25	12.16	1.44	2.05	2.39	5.19	1.90	1.98	2.68	3.48
	Average	NE	3.87	7.39	2.64	3.87	8.45	4.10	15.14	1.36	2.73	2.75	6.49	2.44	2.71	3.73	4.38
	Order	NE	14	24	5	13	25	16	30	2	8	9	20	4	7	11	17

Variables		Variables combinations															
		17	18	19	20	21	22	23	24	25	26	27	28	29	30	31	32
Humidity		yes	yes	yes	yes	yes	yes	yes	yes	yes	yes	yes	yes	yes	yes	yes	yes
Temperature		no	no	no	no	no	no	no	no	yes	yes	yes	yes	yes	yes	yes	yes
# of households		no	no	no	no	yes	yes	yes	yes	no	no	no	no	yes	yes	yes	yes
Population		no	no	yes	yes	no	no	yes	yes	no	no	yes	yes	no	no	yes	yes
Elderly pop.		no	yes	no	yes	no	yes	no	yes	no	yes	no	yes	no	yes	no	yes
Jan.	CV	7.67	8.94	3.98	1.44	8.61	7.58	3.25	1.62	3.62	3.64	8.95	2.93	3.31	3.13	3.08	4.13
	MBE	7.68	9.25	3.92	0.21	8.80	7.81	3.60	0.26	2.75	2.46	9.49	2.38	2.15	1.55	1.92	2.07
	MAPE	5.82	7.01	3.05	1.12	6.67	5.92	2.69	1.12	2.85	2.40	7.19	2.08	2.03	2.33	2.20	2.80
	Average	7.06	8.40	3.65	0.92	8.03	7.10	3.18	1.00	3.07	2.83	8.54	2.46	2.50	2.34	2.40	3.00
	Order	27	30	17	1	29	28	16	2	15	12	31	9	10	7	8	14
Feb.	CV	4.75	6.16	6.84	3.06	7.88	8.37	4.36	11.60	4.62	6.12	9.05	8.57	7.71	8.74	8.15	2.13
	MBE	4.93	6.51	7.35	2.95	8.45	9.02	4.12	12.90	5.08	6.55	9.75	9.58	8.20	9.64	9.11	2.11
	MAPE	3.72	4.91	5.53	2.23	6.36	6.79	3.12	9.70	3.82	4.93	7.35	7.20	6.17	7.24	6.85	1.59
	Average	4.47	5.86	6.57	2.74	7.56	8.06	3.87	11.40	4.50	5.87	8.72	8.45	7.36	8.54	8.04	1.94
	Order	9	12	17	4	20	23	7	30	10	13	27	24	18	25	22	1
Mar.	CV	5.69	6.32	11.34	8.68	8.42	7.66	6.96	15.53	3.89	7.18	6.26	19.17	7.86	15.68	14.57	3.60
	MBE	6.41	6.99	12.74	9.69	9.39	8.51	7.63	17.45	1.76	8.01	6.91	21.61	8.80	17.54	16.02	3.31
	MAPE	4.82	5.26	9.56	7.28	7.05	6.40	5.74	13.09	3.03	6.02	5.19	16.21	6.61	13.17	12.03	2.50
	Average	5.64	6.19	11.21	8.55	8.29	7.52	6.78	15.36	2.89	7.07	6.12	19.00	7.76	15.46	14.20	3.13
	Order	4	7	23	18	17	13	10	29	1	11	6	31	14	30	27	2
Apr.	CV	4.13	7.61	10.63	7.55	7.89	7.10	9.84	11.14	4.56	5.42	9.00	9.17	6.40	7.71	12.03	9.30
	MBE	4.02	8.00	11.34	8.12	8.18	7.31	10.42	12.07	4.74	5.84	9.67	9.75	6.86	7.87	12.97	10.01
	MAPE	3.09	6.05	8.55	6.13	6.18	5.53	7.87	9.10	3.58	4.40	7.29	7.36	5.18	5.95	9.78	7.55
	Average	3.75	7.22	10.17	7.27	7.41	6.65	9.38	10.77	4.29	5.22	8.65	8.76	6.15	7.17	11.59	8.96
	Order	2	13	25	14	16	10	23	26	3	4	19	20	8	12	28	22
May	CV	6.87	4.07	5.83	5.69	5.33	5.59	6.55	9.80	4.90	5.15	12.87	7.74	5.64	7.25	9.60	6.08
	MBE	7.39	3.91	6.10	5.98	5.29	5.77	6.83	10.68	4.99	5.24	14.22	8.44	5.78	7.64	10.40	6.31
	MAPE	5.57	3.09	4.61	4.51	4.00	4.36	5.15	8.05	3.77	3.96	10.71	6.36	4.36	5.77	7.84	4.76
	Average	6.61	3.69	5.51	5.39	4.87	5.24	6.18	9.51	4.56	4.78	12.60	7.51	5.26	6.89	9.28	5.72
	Order	21	1	17	16	10	12	20	29	6	9	31	25	13	22	28	18
Jun.	CV	2.17	3.36	13.99	5.05	5.73	2.57	6.04	8.88	4.50	4.81	11.07	5.63	5.68	5.32	4.62	4.59
	MBE	1.04	2.91	15.53	5.39	5.78	1.76	6.26	9.45	4.95	5.21	12.00	6.14	6.24	5.78	4.89	4.92
	MAPE	1.71	2.20	11.68	4.06	4.36	1.92	4.72	7.12	3.73	3.92	9.04	4.62	4.70	4.35	3.69	3.71
	Average	1.64	2.82	13.73	4.83	5.29	2.09	5.68	8.48	4.39	4.65	10.71	5.46	5.54	5.15	4.40	4.41
	Order	1	4	31	15	19	2	23	25	10	13	28	20	22	17	11	12
Jul.	CV	8.34	3.97	4.14	6.78	5.65	4.22	8.31	10.13	4.09	3.18	4.82	7.65	2.82	6.72	4.57	4.97
	MBE	7.30	2.58	3.39	7.05	4.95	2.34	8.41	10.43	2.45	1.44	4.38	8.15	0.79	6.40	4.22	3.75
	MAPE	6.51	3.14	3.20	5.32	4.34	3.43	6.35	7.86	3.32	2.27	3.63	6.14	1.91	5.21	3.49	3.99
	Average	7.38	3.23	3.58	6.38	4.98	3.33	7.69	9.47	3.29	2.30	4.28	7.31	1.84	6.11	4.10	4.24
	Order	24	7	11	21	16	9	25	28	8	4	15	23	2	20	13	14
Aug.	CV	7.17	6.48	37.29	35.94	14.46	23.07	35.44	38.30	9.84	9.24	29.13	20.16	16.94	16.04	37.55	14.36
	MBE	7.50	5.43	44.79	43.17	16.28	27.10	42.13	45.80	5.67	11.18	33.56	24.40	20.59	18.52	42.97	17.11
	MAPE	5.12	4.06	30.04	28.98	11.03	18.22	28.24	30.77	5.28	7.51	22.49	16.37	13.79	12.20	28.54	11.36
	Average	6.60	5.32	37.38	36.03	13.92	22.80	35.27	38.29	6.93	9.31	28.39	20.31	17.11	15.58	36.35	14.28
	Order	7	5	30	26	12	21	25	31	8	9	23	17	15	14	27	13
Sep.	CV	10.69	8.52	8.03	6.69	9.29	6.64	7.48	8.08	7.92	6.89	7.55	7.04	7.39	10.18	10.42	7.63
	MBE	8.58	4.91	5.24	2.38	6.32	4.72	2.15	4.26	3.84	2.28	3.30	1.56	3.42	8.51	8.58	2.35
	MAPE	6.76	5.57	5.63	4.87	6.13	4.91	4.85	4.73	4.98	4.77	5.24	4.87	5.27	6.78	6.65	4.97
	Average	8.68	6.33	6.30	4.65	7.25	5.42	4.82	5.69	5.58	4.65	5.36	4.49	5.36	8.49	8.55	4.98
	Order	29	18	17	4	25	10	6	12	11	5	9	3	8	26	28	7
Oct.	CV	5.85	3.82	6.32	3.28	4.24	4.89	1.34	10.42	4.90	5.05	6.40	9.62	4.83	4.95	6.46	3.63
	MBE	6.66	4.14	7.37	3.24	4.73	5.58	0.12	12.37	5.44	5.68	7.35	11.30	5.47	5.47	7.42	4.03
	MAPE	4.46	2.77	4.93	2.17	3.16	3.73	0.95	8.26	3.64	3.80	4.92	7.55	3.66	3.66	4.96	2.70
	Average	5.66	3.58	6.21	2.89	4.04	4.74	0.80	10.35	4.66	4.84	6.22	9.49	4.65	4.69	6.28	3.45
	Order	25	12	26	9	15	19	1	31	17	20	27	30	16	18	29	11
Nov.	CV	10.01	4.55	4.45	6.58	5.57	5.77	5.71	4.51	5.39	7.15	7.11	5.58	6.69	6.56	8.15	4.08
	MBE	10.40	5.28	4.71	7.07	6.09	6.69	6.35	5.30	6.19	7.99	8.34	6.29	7.44	7.24	9.58	4.22
	MAPE	6.96	3.52	3.15	4.73	4.07	4.47	4.24	3.54	4.14	5.34	5.57	4.20	4.97	4.84	6.40	2.82
	Average	9.13	4.45	4.10	6.13	5.24	5.64	5.43	4.45	5.24	6.83	7.01	5.36	6.37	6.21	8.04	3.71
	Order	31	6	5	21	15	19	18	7	14	26	28	17	24	22	30	3
Dec.	CV	4.05	7.33	1.89	12.20	6.33	10.29	12.99	15.54	2.34	3.75	9.41	6.70	3.99	3.89	5.51	6.57
	MBE	4.62	8.70	0.38	14.87	7.41	12.41	15.74	18.64	1.73	2.35	11.45	8.20	2.55	4.59	6.66	8.02
	MAPE	3.09	5.81	1.25	9.92	4.95	8.28	10.50	12.45	1.37	1.94	7.64	5.47	2.23	3.05	4.44	5.35
	Average	3.92	7.28	1.17	12.33	6.23	10.33	13.08	15.54	1.81	2.68	9.50	6.79	2.92	3.84	5.53	6.64
	Order	15	23	1	28	19	27	29	31	3	6	26	22	10	12	18	21

Table 4. Best Combination of Variables and Corresponding Metric Results using Data Sample 2

Month of the Year	Best Average Error Rate	Best Variable Combination (number / combination)
January	0.92	20 / humidity, population, elderly population
	1.00	24 / humidity, households, population, elderly population
	1.26	7 / households, total population
February	1.94	32 / humidity, temperature, households, population, elderly population
	2.55	8 / households, population, elderly population
	2.64	15 / temperature, households, population, elderly population
March	2.89	25 / humidity, temperature
	3.13	32 / humidity, temperature, households, population, elderly population
	5.33	10 / temperature, elderly population
April	2.67	9 / temperature
	3.75	17 / humidity
	4.29	25 / humidity, temperature
May	3.69	18 / humidity, elderly population
	3.95	10 / temperature, elderly population
	4.09	12 / temperature, population, elderly population
June	1.64	17 / humidity
	2.09	22 / humidity, households, elderly population
	2.71	9 / temperature
July	1.16	14 / temperature, households, elderly population
	1.84	29 / humidity, temperature, households
	1.91	16 / humidity
August	3.68	5 / households
	4.09	2 / elderly population
	4.22	9 / temperature
September	3.03	12 / temperature, population, elderly population
	4.27	14 / temperature, households, elderly population
	4.49	28 / humidity, temperature, population, elderly population
October	0.80	23 / humidity, households, population
	0.97	8 / households, population, elderly population
	1.82	10 / temperature, elderly population
November	0.86	4 / population, elderly population
	1.25	7 / households, total population
	3.71	32 / humidity, temperature, households, population, elderly population
December	1.17	19 / humidity, population
	1.36	9 / temperature
	1.81	25 / humidity, temperature

IOP Conf. Series: Journal of Physics: Conf. Series **1072** (2018) 012005 doi:10.1088/1742-6596/1072/1/012005

Table 5. Frequency of Variables in Combinations with Lowest Error Rate

Month of the year	Variable	First 6 Variable Combinations with Lowest Error Rate						Frequency of variable
		1	2	3	4	5	6	
January	Humidity	yes	yes	no	no	no	no	2
	Temperature	no	no	no	yes	no	yes	2
	# of households	no	yes	yes	no	no	no	2
	Population	yes	yes	yes	yes	yes	no	5
	Elderly population	yes	yes	no	yes	yes	yes	5
February	Humidity	yes	no	no	yes	no	no	2
	Temperature	yes	no	yes	no	no	yes	3
	# of households	yes	yes	yes	no	no	no	3
	Population	yes	yes	yes	yes	yes	no	5
	Elderly population	yes	yes	no	yes	no	no	3
March	Humidity	yes	yes	no	yes	no	yes	4
	Temperature	yes	yes	yes	no	no	yes	4
	# of households	no	yes	no	no	no	no	1
	Population	no	yes	no	no	no	yes	2
	Elderly population	no	yes	yes	no	yes	no	3
April	Humidity	no	yes	yes	yes	no	no	3
	Temperature	yes	no	yes	yes	no	yes	4
	# of households	no	no	no	no	no	no	0
	Population	no	no	no	no	no	no	0
	Elderly population	no	no	no	yes	yes	yes	3
May	Humidity	yes	no	no	no	no	yes	2
	Temperature	no	yes	yes	yes	no	yes	4
	# of households	no	no	no	yes	yes	no	2
	Population	no	no	yes	yes	yes	no	3
	Elderly population	yes	yes	yes	no	yes	no	4
June	Humidity	yes	yes	no	yes	no	no	3
	Temperature	no	no	yes	no	no	yes	2
	# of households	no	yes	no	no	no	no	1
	Population	no	no	no	no	no	yes	1
	Elderly population	no	yes	no	yes	yes	yes	4
July	Humidity	no	yes	no	yes	no	no	2
	Temperature	yes	yes	yes	yes	yes	yes	6
	# of households	yes	yes	yes	no	yes	no	4
	Population	no	no	yes	no	yes	no	2
	Elderly population	yes	no	yes	yes	no	yes	4
August	Humidity	no	no	no	no	yes	no	1
	Temperature	no	no	yes	yes	no	yes	3
	# of households	yes	no	no	no	no	yes	2
	Population	no	no	no	no	no	yes	1
	Elderly population	no	yes	no	yes	yes	no	3
September	Humidity	no	no	yes	yes	yes	yes	4
	Temperature	yes	yes	yes	no	yes	no	4
	# of households	no	yes	no	no	no	yes	2
	Population	yes	no	yes	yes	no	yes	4
	Elderly population	yes	yes	yes	yes	yes	no	5
October	Humidity	yes	no	no	no	no	no	1
	Temperature	no	no	yes	no	no	yes	2
	# of households	yes	yes	no	yes	yes	yes	5
	Population	yes	yes	no	no	yes	no	3
	Elderly population	no	yes	yes	yes	no	no	3
November	Humidity	no	no	yes	no	yes	yes	3
	Temperature	no	no	yes	yes	no	no	2
	# of households	no	yes	yes	no	no	no	2
	Population	yes	yes	yes	yes	yes	no	5
	Elderly population	yes	no	yes	yes	no	yes	4
December	Humidity	yes	no	yes	no	no	yes	3
	Temperature	no	yes	yes	yes	no	yes	4
	# of households	no	no	no	yes	no	no	1
	Population	yes	no	no	no	yes	no	2
	Elderly population	no	no	no	no	yes	yes	2
All months	Humidity	yes	no	yes	no	no	yes	30
	Temperature	no	yes	yes	yes	no	yes	40
	# of households	no	no	no	yes	no	no	25
	Population	yes	no	no	no	yes	no	33
	Elderly population	no	no	no	no	yes	yes	43

IOP Conf. Series: Journal of Physics: Conf. Series **1072** (2018) 012006 doi:10.1088/1742-6596/1072/1/012006

THE Simulation Study of DC Grounding Electrode Based on CDEGS and ANSYS

Chenglian Ma[1], Shujian Zhao[1], Li Sun[1]*, Letian Wang[2] and Guangming Zhang[2]

[1] School of Electrical Engineering, Northeast Electric Power University, Jilin, Jilin 132012, China

[2] ShanDong Electric Power Engineering Consulting Institute CORP.,LTD. Jinan, Shandong 250013, China

*Li Sun: sunlinedu@163.com

Abstract. When the UHVDC project was put into operation, grounding electrode grounding current may affect the transformer in the substation near the pole, it will cause transformer DC bias saturation, which leads to higher transformer temperature rise, too high harmonic and reactive power consumption, which will cause transformer damage or power grid blackout. The earth resistivity model of grounding pole is not accurate enough and the calculation of surface potential distribution is not accurate enough, which will cause the deviation between the evaluation and actual damage degree before and after the project put into operation. In order to simulate the grounding potential distribution directly after grounding, the simulation analysis with different methods can be checked with each other, and the related design units used CDEGS to simulate the earth potential rise. In this paper, the influence of DC transmission grounding electrode on the distribution of ground surface potential is analyzed, and the calculation method of DC grounding electrode current field is deduced. The 6 layers earth soil model was established. CDEGS and ANSYS were applied. The earth potential rise distribution in the 0~100 km radius of the earth electrode was calculated respectively. In this paper, the results of the earth potential distribution in the 0-50 km range of the two schemes were compared and analyzed. The results show that the calculation results of ANSYS were slightly smaller than those calculated by CDEGS, and the difference was about 0.6 V. The calculation results of the two schemes could provide reference for field practical monopole operation and commissioning.

1. Introduction

The development of ultra high voltage AC / DC power transmission is in line with national conditions, and it is an important measure to prevent and control air pollution. The HVDC project is designed to form a circuit and set a potential reference point. It is necessary to set the DC ground electrode [1]. In the initial stage of completion and the operation mode transformation phase during the operation process, the operation mode of monopole earth loop will be used. Up to thousands of ampere currents flow into the earth through the earth's pole. And generally it will last for a long time at the ground pole. A series of hazards may be induced [2].

With the construction of UHVDC converter station, it is revealed that the grounding electrode has more and more influence on the surrounding AC power system. The normal alternating current magnetic field of a power transformer is biased by the impact of the intrusions of DC current. The power transformers with DC bias become the harmonic power and reactive power of the AC power

IOP Conf. Series: Journal of Physics: Conf. Series **1072** (2018) 012006 doi:10.1088/1742-6596/1072/1/012006

grid. A large number of harmonic currents can cause the relay protection to be misoperated or rejected. The phenomenon of a large number of transformers in the power grid with DC bias will cause the system's reactive power fluctuation and voltage drop. When the phenomenon of DC bias is serious, it can cause permanent damage to the transformer body. Therefore, the analysis and grasp of the DC bias of the power transformer is also the basis for the rational assessment of the influence of DC bias and the necessary measures to be taken [3-4].

The key problem at the design stage is the accuracy of the geodetic potential distribution around the ground polar site. This problem plays an important role in evaluating the direct current derived disasters. This paper provides a certain theoretical reference for the calculation of the ground potential distribution in HVDC transmission project. Combined with the earth pole project of the Northern Shaanxi converter station, the soil resistivity model of multi-layer soil near ground pole is simulated by using CDEGS and ANSYS. The simulation results show the distribution of the surface potential near the ground pole. The CDEGS calculation data and the ANSYS simulation data are compared and analyzed. The research work depends on the grounding pole engineering of the existing North Shaanxi ±800 kV UHVDC converter station, and carries out the calculation and study of the grounding electrode. The calculation is carried out in this paper so that the designer can fully grasp the design scheme of the grounding electrode and the calculation of the influence of the earth potential rise.

2. Basic Equation and Boundary Condition of Geoelectric Field

2.1. Basic Equation

When the DC transmission lines operate in the monopole circuit and the bipolar unbalance earth return line operation, the current is in the way of the earth, and a direct current potential will be generated on the surface of the earth [5]. The structure of the monopole earth line and the structure diagram of the operation mode of the bipolar unbalance earth line are as shown in Figure 1 and Figure 2:

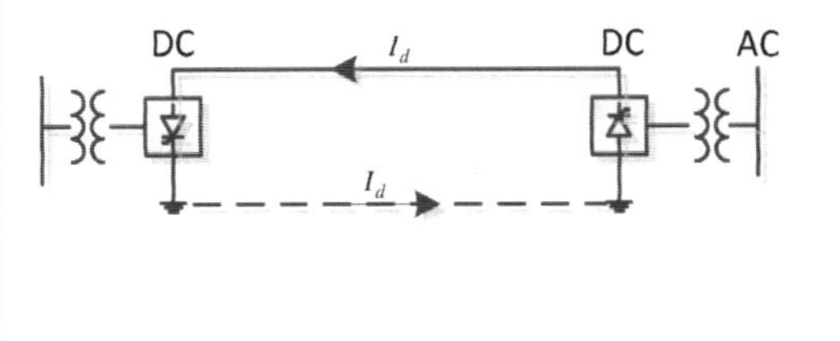

Figure 1. Ground return operation mode of HVDC transmission system Monopole ground return model.

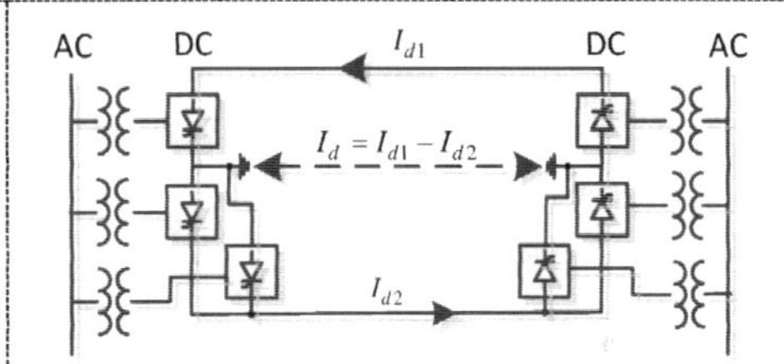

Figure 2. Ground return operation mode of HVDC transmission system Bipolar unbalanced earth loop operation.

2.2. Boundary Condition

Assuming that a DC grounding current I nearby the grounding electrode. In the ground 5 meters underground within the scope of the regional point current source is I, the unit point charge density function is defined as a formula (1).

$$\rho(x) = \delta(x - x') \tag{1}$$

where, $\delta(x - x')$ is defined as the unit point charge density, at the point x'.

$$\begin{cases} \delta(x) = 0 & x \neq 0 \\ \int_v \delta(x) dv = 1 & \text{else} \end{cases} \tag{2}$$

It can be seen that the point charge is the limit of a charged ball with a small volume and a large charge density, distributed within a small volume ΔV. When $\Delta V \to 0$, $\rho \to \infty$. The values of $\nabla \cdot \vec{J}$ with current source are derived by formula (3) and formula (4):

$$I = \oint_s \vec{J} \cdot d\vec{s} = \int_\Omega \nabla \cdot \vec{J} d\Omega \tag{3}$$

$$I = I \times 1 = \int_\Omega I\delta(x) d\Omega \tag{4}$$

3. MT Measurement of Earth Resistivity of Grounding Electrode in North Shaanxi Converter Station

3.1. Engineering Survey

The rated transmission capacity of 10000 MW is ± 800 kV Shaanxi Lake UHVDC transmission project. The connection of the valve group of the rectifier station is a bipolar connection. The connection of the valve group of the inverter station is also a bipolar connection. And the double 12 pulsating valve group is used for each connection. The sending end station grounding engineering is an important part of Shaanxi Lake UHVDC project.

According to the information of the system, the construction scale of the transfer station of this project is as follows: The DC current is rated 6250 A, and the transmission capacity is rated 10000 MW. According to the recommendation of the site selection of the transfer station in the feasibility study report of the Shaanxi Lake project, the sending end station grounding electrode site recommends alternative for Fugu site. This project is located in the Loess Plateau. The landform is gently sloping and sloping. The terrain breakage unit is complex. On the whole, the soil structure and its topography are more complex.

3.2. Ground Pole Soil Parameters

In this project, the resistivity of shallow soil has been surveyed in the research stage. In addition to conventional resistivity measurements, the MT is also used to detect the deep electrical structure characteristics of the recommended polar site. After the systematic analysis and processing of the interpretation results of magnetotelluric sounding data, the electrical parameters have been obtained about Yan site of Daliushu village, the multi-layer soil model is shown in Figure 3.

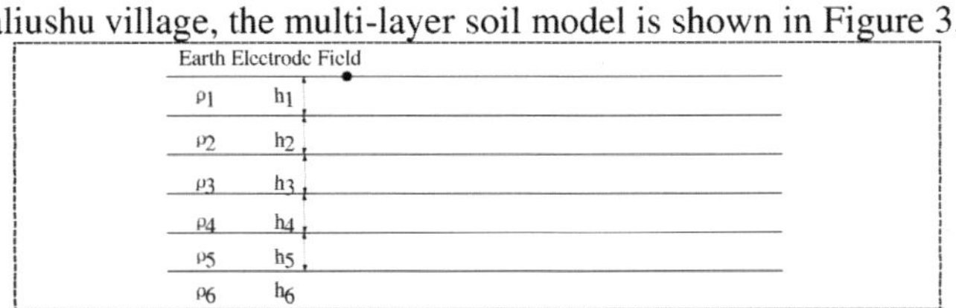

Figure 3. Model of multi-layer soil.

According to the surface soil resistivity and the deep soil resistivity measured by magnetotelluric method, the surface soil resistivity and the magnetotelluric method are measured. The soil structure model near the ground pole site was studied. On the basis of determining the accurate soil model, ANSYS finite element analysis software and CDEGS software are applied to model the resistivity structure around the pole. The simulation calculation is carried out. The potential rise of various substations is obtained. Based on the CDEGS software and the ANSYS software, the design department can be calculated and analyzed.

4. Results and Comparison

4.1. Calculation Results by CDEGS

The geodetic potential of the earthing pole is up to 532 V. Geodetic potential of the 16km rises to less than 6.9 V. The potential of the 33km rises to less than 3 V.

It can be seen from the Fugu Yan village site of Daliushu substation and power plant within 50 km, except the 330 kV Hao Jia substation, the current of the other main transformer core does not exceed the allowable range.

4.2. Calculation Results by ANSYS

The earth potential rise data curve in the 0~100 km range of the ground electrode in Figure 4, and the 0~50 km range earth potential rise curve is shown in Figure 5. The 0~10 km range earth potential rise data curve is calculated by ANSYS as shown in Figure 6, and the 0~5 km range earth potential rise curve is shown in Figure 7.

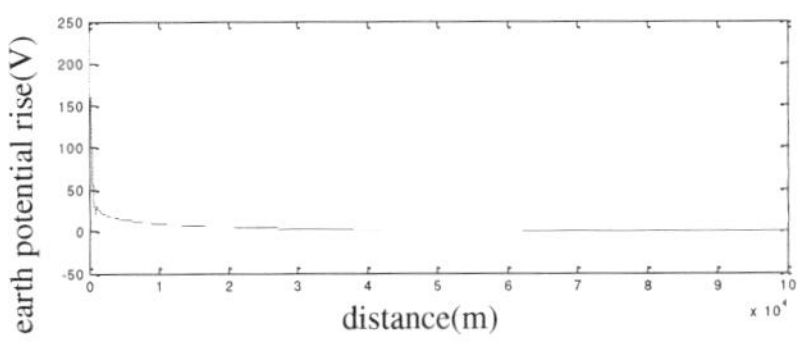

Figure 4. Earth potential rise data curve in the 0~100 km range of the ground electrode.

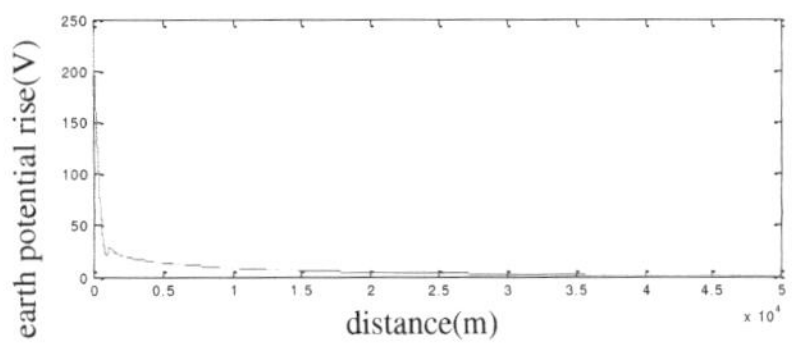

Figure 5. Earth potential rise data curve in the 0~50 km range of the ground electrode.

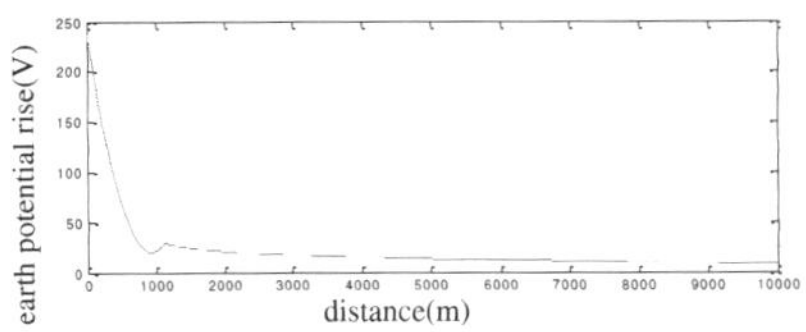

Figure 6. Earth potential rise data curve in the 0~10 km range of the ground electrode.

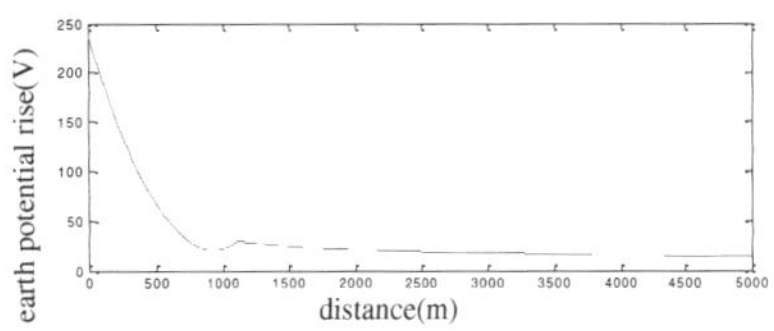

Figure 7. Earth potential rise data curve in the 0~5 km range of the ground electrode.

Through ANSYS calculation and analysis, in conjunction with Fig. 4, Fig. 5, Fig. 6, fig 7, the geodetic potential rise curve within the range of 100 km of the polar address of the ground pole is connected to the range of the polar address of the ground pole. It can be seen that the trend of the geodetic potential rise is decreasing. The fastest range of reduction is 2 km of the earth's polar square. The problem of ground potential distribution in the 2 km range of the earth's polar square should be taken into consideration. The effect of the ground potential is faster and the ground potential gradient is large, and the effect on the ground artificial system is greater. The land into the earth is deep in the depths of the earth. The current density on the surface gradually decreases as it is far away from the ground pole.

4.3. Results Comparison

The comparison between the results of ANSYS software and the results of CDEGS can be seen. The calculated results of ANSYS are slightly less than the CDEGS results, and the difference is about 0.6 V. According to the results of this calculation, the potential difference between the nodes calculated by CDEGS and ANSYS is basically no difference. Therefore, the application of electric network theory and the analysis of node voltage method can be found. The calculated current value of the DC current flowing through the neutral points of each transformer should be basically the same.

The result of ANSYS calculation is slightly lower than that of CDEGS. The authors believe that the ANSYS finite element analysis software is used to calculate the distribution of the geodetic potential in large scale structures of the earth. The CDEGS software package calculates the geodetic potential rise, and the results of the ANSYS calculation are slightly lower than the results of the CDEGS software package.

However, the real geodetic structure is more complex. In this paper, the 6 layer earth resistivity data of magnetotelluric sounding in the field of MT method is adopted by the relevant design institutes. The model is built on horizontal uniform layered structure. The effects of actual breakage, lateral difference and coastal effect are not considered. So this paper simulates the basic data used in simulation. There is a great difference from the actual geoelectric structure. The 0-5 km range geodetic elevation data curve of the distance to the ground pole address of Figure 7 shows that, when the distance to the ground pole is above 2 km, the decrease of the surface potential slows down. And we should focus on the surface potential distribution of the ground polar square 2 km, as described in the previous article. The construction of related power facilities and oil and gas pipelines should be avoided in the square circle. This result can also provide reference for the initial site selection of DC transmission project, and away from the related facilities in the near area. At the same time, the damage prevention and control of the built power facilities and oil and gas pipelines should be strengthened.

5. Conclusion

(1) For the surface potential distribution, in this paper, the 6 layer soil resistivity model measured by MT method is used to calculate the CDEGS software and the ANSYS software. The calculation results show that the distance from the ground to ground electrode decreases gradually. And two methods are both low and fast in the nearer ground potential.

(2) The results are calculated according to the ground potential rise of two kinds of software. The difference between the potential differences between the nodes calculated by CDEGS and ANSYS is smaller. Therefore, the application of electrical network theory analysis shows that The two kinds of software calculated by the neutral point of each transformer should be basically the same.

6. References

[1] Liu Lianguang, Yan Xudong, Ma Chenglian, Wei Kai and Zhang Yanbing 2016 *Research of converter transformer marshalling and receiving-end grid structure's effect on converter transformer DC bias* (Power System Technology)

[2] Mousavi S A and Bonmann D. 2017 *Analysis of asymmetric magnetization current and reactive power demand of power transformers due to GIC* (Procedia Engineering) **202** 264-272

[3] Kazerooni M, Zhu H and Overbye T J 2016 *Improved Modeling of Geomagnetically Induced Currents Utilizing Derivation Techniques for Substation Grounding Resistance* (IEEE Transactions on Power Delivery) p99

[4] Babazadeh D, Muthukrishnan A, Mitra P, Larsson T and Nordström L 2017 *Selection of dc voltage controlling station in an HVDC grid* (Electric Power Systems Research) **144** 224- 232

[5] Shetye K and Overbye T 2015 *Modeling and Analysis of GMD Effects on Power Systems: An overview of the impact on large-scale power systems* (IEEE Electrification Magazine) vol 3 chapter 4 pp 13-21

Acknowledgments

This work was supported by National Natural Science Foundation of China(51577060, 51677068); The National High Technology Research and Development of China(863 Program)(2012AA121005). The National Key Research and Development Program of China (No. 2017YFB0902000).

IOP Conf. Series: Journal of Physics: Conf. Series **1072** (2018) 012007 doi:10.1088/1742-6596/1072/1/012007

THE Study on the Polar Potential Distribution of DC Grounding in Earth Structures

Chenglian Ma[1], Shujian Zhao[1], Li Sun[1]*, Letian Wang[2] and Tao Wang[2]

[1] School of Electrical Engineering, Northeast Electric Power University, Jilin, Jilin 132012, China

[2] ShanDong Electric Power Engineering Consulting Institute CORP.,LTD. Jinan, Shandong 250013, China

* Li Sun: sunlinedu@163.com

Abstract. This paper introduced the boundary conditions of the complex soil potential. Then the simulation is carried out with the ANSYS software. And effect of the ground potential distribution in rivers and mountains of the complex geological conditions are analyzed. It is pointed out that rivers can change the potential distribution, and mountains can hardly change the potential. The rational use of these characteristics can effectively alleviate the effect of the grounding electrode potential on the power system and the environment. At last, the importance of establishing a three-dimensional geoelectric field model is put forward, and the geodetic structure is more accurately understood by considering the transversal difference of the earth. Then the current field distribution around the DC ground electrode is calculated accurately. It is better to serve the construction of UHVDC power transmission project in China.

1. Introduction

The initial and maintenance stages of the UHVDC power transmission project in China are completed. It is usually temporarily operated in a monopole circuit, there will be a strong DC current (3000A, 5000A, and 6250A) at this time. For a long time into the earth, potential distortion is caused in a wide range of ground. It will be around the environment, the adverse effects of oil gas pipeline network and power grid communication system are harmful. The key to solve the above problems lies in exploring the flow law of current in the soil and the distribution of the ground potential [1].

There have been many scholars in the field of geodetic potential analysis. In most previous analyses of the geodetic potential, the commonly used soil model is the horizontal homogeneous multi-layer soil structure [2-4]. There is no analysis of the influence of geodetic structure on the surface potential distribution. Therefore, the influence of the structure of the soil on the neutral current of the transformer can not be obtained from the results of the analysis. The soil structure is actually uneven. On the macro level, there are horizontal and vertical partitioning. As a mountain area, the soil structure will be more complex in the near sea or lake area [5].

In this paper, the finite element method is applied to the calculation of the ground current field. The calculation formula of the DC ground potential under the complex geodetic structure is derived. Then the simulation analysis is carried out by using ANSYS. The distribution of surface potential in the presence of rivers and mountains is simulated. The influence of geodetic structure on the distribution of ground potential is analyzed. The distribution of surface potential in the unipolar operation of DC grounding electrode with different influence factors is summarized. It provides reference for the design and construction of the later UHVDC transmission project.

2. The Influence of Geodetic Structure on the Distribution of Ground Potential

The current of the rated operating ±800 kV UHVDC power transmission projects, which have been put into operation, is 3000 A or 4000 A. The current of the ZhunDong - Wannan ±1100 kV UHVDC in the construction is 5000 A. And the construction scale of the transmission terminal of the Hubei ±800 kV UHVDC transmission project as follow: the transmission capacity of the DC system is 10000 MW, and the rated DC current is 6250 A.

A single phase transformer of ZZDFPZ-321000/500 model is adopted for the change and rheology of ±800 kV. Through the calculation of the nameplate parameters, the DC resistance of the windings is 0.704, and the grounding resistance of the converter station is 0.1. ±500 kV AC substations usually use single-phase autotransformer. The DC resistance of high voltage and medium voltage windings is 0.238 and 0.097, respectively. The grounding resistance of 500 kV substations is 0.2. The AC lines use four split wires. The resistance of single phase line is 0.0187 /km.

It can be seen that the DC resistance of HVDC transmission lines is small. For an AC grid, if there is a potential difference between two points on the ground, and there is an electrical connection between the ground and the ground system, the grounding of the ground pole to the ground will flow into the power grid, which has a serious impact on the power grid system, resulting the DC bias of the AC transformer, vibration and noise, and the load increased.

However, the vast majority of the currents flow through the earth. Therefore, it is necessary to analyze the impact of the direct current on the power system or the environment. The key problem is to analyze the distribution of the geodetic soil in the earth and the potential distribution on the ground. From some of the existing DC projects that have been put into operation, the influence of DC transmission to the polar potential is often underestimated, even the wrong estimate. The distribution of the ground potential is not only determined by the distance from the grounding electrode. It is related to the stratification of the soil and the distribution of the surrounding rivers, lakes, seas and mountains, especially for the study of large range potential distribution.

Considering the effect of geodetic structure on the distribution of ground potential is not only used in HVDC. It can also provide reference for the study of the phenomenon of high current to ground such as lightning. On the other hand, it has value for the study of the distribution of magnetic induction current GICs on the earth, ESP and so on.

3. Analysis Method and Boundary Condition Setting

3.1. Analysis Method

With the help of ANSYS, the simulation of the ground potential under different geodetic conditions is completed. Under the complex geodetic structure, if the traditional mirror method or the complex image method is used to calculate, it is inevitable that a large number of formula derivation should be carried out. The earth is not always horizontal or vertically homogeneous. The derivation of different ground potential formulas for various complex terrains is not only tedious but also inefficient. The most suitable method is the method of field analysis. The equation of different topographic field is constant. As long as the corresponding boundary conditions are set for the interface between different terrain, it is possible to use the method of numerical calculation to solve the problem quickly. At the same time, the application of the theory and research method to the finite element calculation of the three dimensional electromagnetic field of the transformer is also a subject worthy of study.

3.2. Boundary Condition Setting

The distribution of ground potential and the striding voltage and grounding resistance in the vicinity of the ground electrode are analyzed. Then a model of soil and grounding electrode near the pole site is established. An axisymmetric model is established in ANSYS. Using current field calculation module, the unknown field is the potential value V in the area of the 0 reference point at infinity. At the same time, the soil resistivity of each layer is defined. The vertical external method of the surface potential is 0. The underlying potential is set to 0. The choice of excitation is determined by the method of defining the potential value at the ground pole. First, the voltage on the grounding electrode is defined. Then the finite element calculation is carried out. Calculate the electric field strength of each side of the grounding electrode E,

$$E = 2\pi R\int E_x dx$$
$$E = 2\pi R\int E_y dx \qquad (1)$$

Then the grounding electrode current I is obtained.

$$I = E / \rho \qquad (2)$$

where, R is ground pole radius, Ex is The electric field intensity (V/M) in the radial direction of the electrode, Ey is the electric field intensity (V/M) in the vertical direction of the electrode profile, ρ is the soil surface (Ω - M) of electrical resistivity.

4. Simulation Results and Analysis

The distribution of the ground potential is calculated in the following parts of the sea and the high mountain. The point current source is used. The range is from a radius of 50 to 100 km. The soil model is illustrated by the typical tectonic model of the earth, as shown in Table 1. The effect of geodetic structure on the ground potential is analyzed in the light of the calculation results.

Table 1. The caption must be followed by the table.

	Thickness of layer (km)	resistivity (Ω•m)
ρ0	0.1	10
ρ1	2	200
ρ2	30	10000
ρ3	100	0

4.1. The Influence of Rivers

In this study, two river models are established. One is perpendicular to the radius of the earth's polar radius, and the other is along the radius of the earth's pole. For the first case, a soil model with long and wide 100 km and 100 km long rectangles is established. The river model is a small rectangular body with a vertical distance of 20 km, a width of 1 km, a depth of 1 km, and a long 100 km on the surface of the soil. The conductivity of the river is 1 μ S/cm. For the second case, the same soil model was established to solve the lake. The river model is taken from 5 km to 40 km from the central point, 1 km in width and 1 km in depth on the surface of the soil.

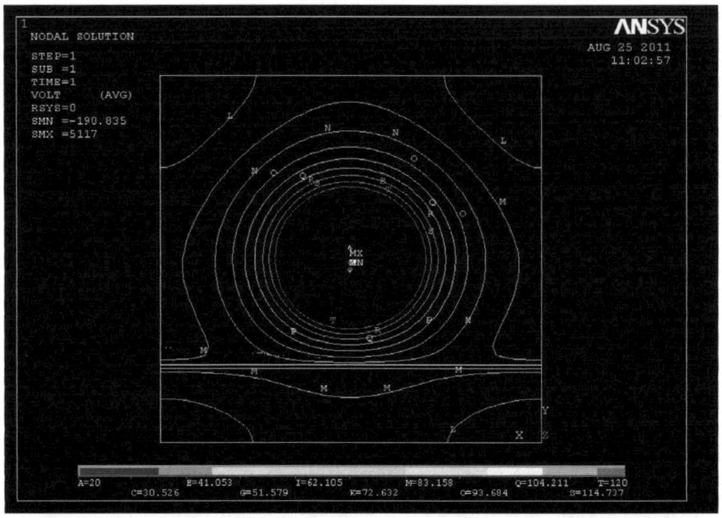

Figure 1. The distribution of the potential of the lateral rivers.

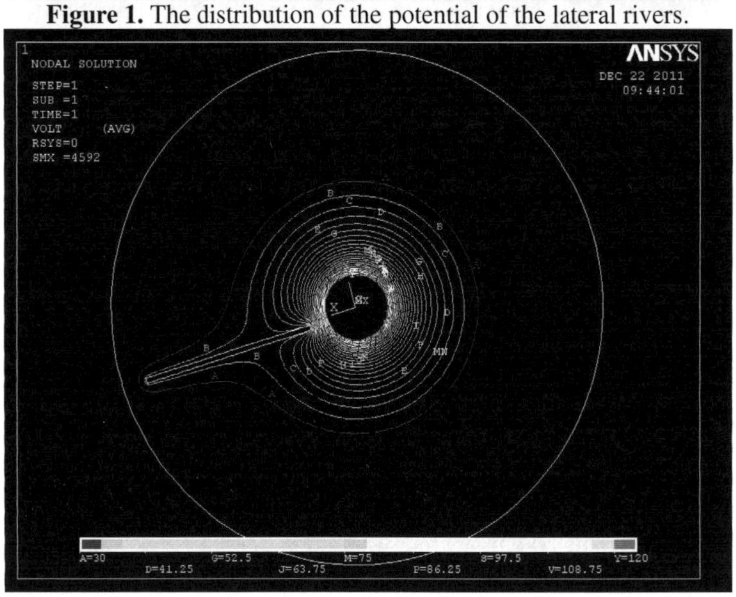

Figure 2. The distribution of potential distribution of longitudinal rivers.

Figure 1 is the distribution of the equipotential line of the earth when the lateral river exists in the vertical direction of the earth's radius. As shown in Figure 1, the direction of the river is parallel to the direction of the equipotential line. Therefore, the lateral river has little effect on the geodetic potential. The entire surface of the river is equipotential, and the decrease of the geodetic potential of the river at the ground pole is slightly quicker. On the outside of the river, the decrease of the potential of the earth slightly slows down. In general, the impact is not very big.

Figure 2 is the direction of the flow of a river along the radius of the earth's electrode. That is, the distribution of the geodetic potential in the presence of a longitudinal river. As shown in Figure 2, the change of potential in the area flowing through the river is obvious. Firstly, in the extreme part of the river near the ground, the potential changes sharply. The partial potential change in the river far away from the ground pole is slow. Secondly, the river at this time is similar to a wire that flows through the current, and the resistance is small, and the potential is small. The whole river spans tens of kilometers in the direction of the radius. The whole river spans tens of kilometers in the direction of the radius. But the whole potential drop is less than 5 V. It also affects the whole river area, and the rate of potential decline also slows.

In the presence of a river that flows along a radius in the vicinity of the DC ground, one is to consider the problem of the large current distribution density of the earth's surface near the extreme near the ground. In addition, if the conditions are available, the average potential action of the river can be used to reduce the effect of the ground potential difference on the power system and the environment.

4.2. The Influence of Mountains

In this chapter, the surface potential is simulated with the distribution of mountain ranges around the ground pole. The soil model uses the typical soil model of Table 1. The alpine model uses a cone. The surface diameter of the upper surface is 1 km, the lower surface diameter is 20 km, and the height is 10 km. The center of the mountain is located at 70 km from the ground to the ground. The mountains are basically composed of rocks. In this paper, 1000 Omega m is taken.

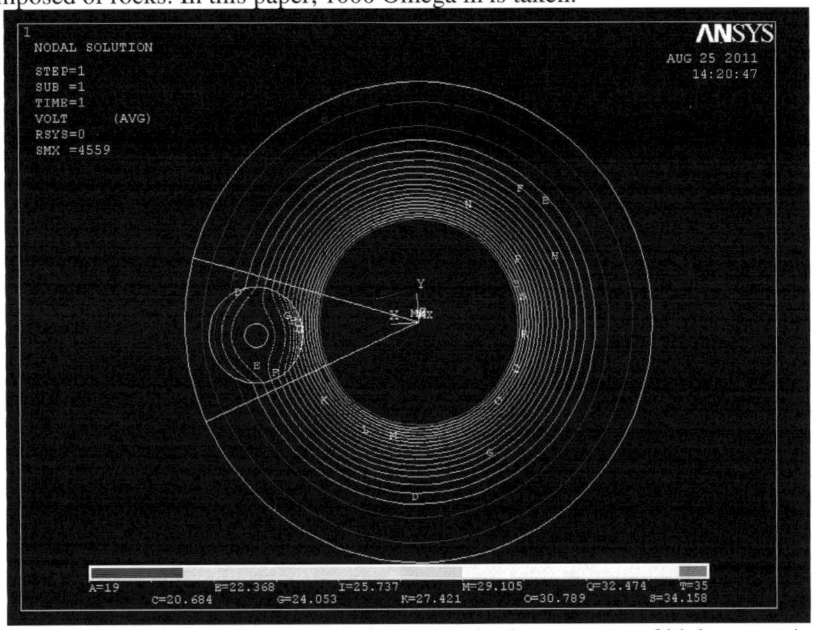

Figure 3. The earth surface potential top view in the presence of high mountains.

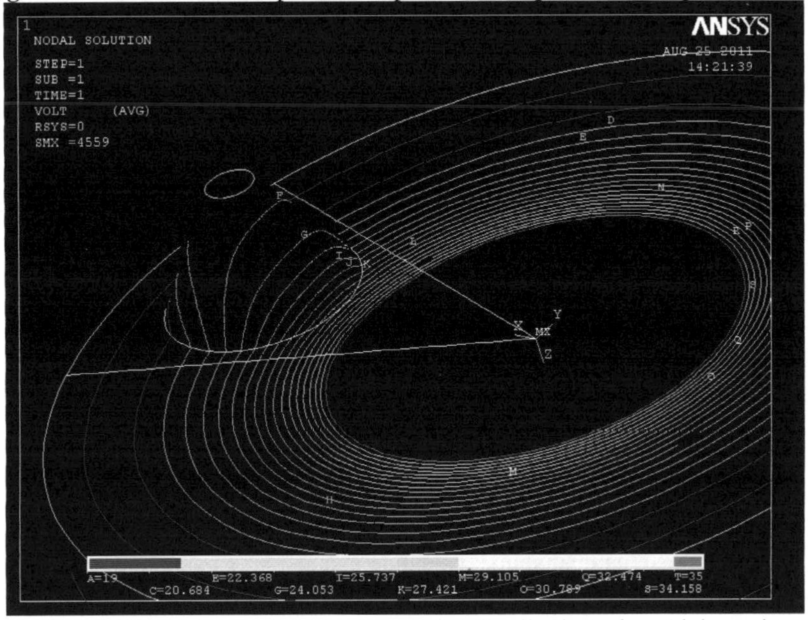

Figure 4. Lateral view of the earth potential distribution when Alpine exists.

Figure 3 is a view of the potential distribution of the earth's surface potential in the presence of Alpine. It can be seen that the geodetic potential of the region in which the mountain is located is distorted. But in the area outside the mountains, the potential is distributed according to the normal concentric circle. It's basically unaffected by the mountains.

Figure 4 is a view of the side potential distribution in the presence of a high mountain. Because the resistivity of the mountain is higher than that of the ordinary soil, in theory, when the current flows through the mountains, the potential drops rapidly. However, in figure 4, after the current flows through the mountains, the absolute value of the potential at the other end of the mountain is not the same as that in the normal condition. And the potential distribution on the mountain is thiner, and the potential drop is slower than that in the normal condition. This is due to the high resistivity of the high mountain and the low flow of current. On the other hand, the height of the mountains is large, the surface area is large, and the area of the current flows is larger. These two aspects work together, and the flow of current is just parallel to the surface of the potential line on the surface of the surrounding ground. In the rear of the mountain, the distribution of potential is the same as that under normal conditions.

According to the above situation, when there is a high mountain near the ground pole, it is unnecessary to think about the impact of high mountains. At the time of calculation, the mountain can be ignored and the calculation can be simplified.

5. Conclusion

(1) The earth potential equation and boundary condition are derived under the condition of complex soil structure. The finite element method is applied to solve the problem. A suitable model can be set up, and the boundary conditions and the source of the field are set up scientifically. With the help of the finite element ANSYS, the distribution of the ground potential is obtained.

(2) Because the structure of the earth is complex, the range of geodetic measurements in the site of the grounding electrode is limited. There is a great deviation between the actual situation of the grounding current distribution of the grounding electrode and the theoretical analysis and calculation. The necessity of establishing three-dimensional geodetic resistivity model in large area is becoming more and more important.

6. References

[1] Li H, Cui X, Liu D, Lu T and Cheng Z 2010 *Influence on Three-Phase Power Transformer by DC Bias Excitation* (Transactions of China electro technical society), vol 25 chapter 5 pp 88-96

[2] Kazerooni M, Zhu H, Overbye T and Wojtczak D 2016 *Transmission system geomagnetically induced current model validation* (IEEE Transactions on Power Systems), p 99

[3] Mousavi S A and Bonmann D 2017 *Analysis of asymmetric magnetization current and reactive power demand of power transformers due to GIC* (Procedia Engineering) **202** 264-272

[4] Kazerooni M, Zhu H and Overbye T J 2016 *Improved Modeling of Geomagnetically Induced Currents Utilizing Derivation Techniques for Substation Grounding Resistance* (IEEE Transactions on Power Delivery) p99

[5] Hussein A A and Ali M H 2017 *Fuzzy logic controlled variable resistor for suppressing GIC in transformers*(Iet Generation Transmission & Distribution) vol 11 chapter 6 pp 1494-1501

[6] Rezaei-Zare A 2015 *Reactive Power Loss Versus GIC Characteristic of Single-Phase Transformers*(IEEE Transactions on Power Delivery) vol 30 chapter 3 pp 1639-1640

[7] Kovan B and De Leon F 2016 *Mitigation of Geomagnetically Induced Currents by Neutral Switching*(Power Delivery IEEE Transactions on) vol 30 chapter 4 pp 1999-2006

Acknowledgments

This work was supported by National Natural Science Foundation of China(51577060, 51677068); The National High Technology Research and Development of China(863 Program)(2012AA121005). The National Key Research and Development Program of China (No. 2017YFB0902000).

Study on the Control System of the Phase Controlled Rectifier for Mine

Quanzhu Zhang, Haiyu Chen, Yuanhong Li and Yonghong Deng

North China Institute of Science and Technology, China

qzhzhang@ncist.edu.cn; 1195078260@qq.com; dyhsyjdyx@163.com

Abstract. In order to meet the requirements of different voltage rating of different electric equipment in coal mine and the requirement of voltage stabilizing for equipment use. A new type of phase control rectifier control system for mine is studied in this paper. The system has designed the control system based on the chip DSP software and the closed loop control system of the software PID. The experimental results show that the phase controlled rectifier control system can achieve a good phase shifting regulation for the rectifier output voltage, and control the output voltage of the trigger angle to meet the needs of the rated voltage of different electrical equipment, which is flexible and convenient. The software PID can adjust the closed loop according to the set value of voltage and current to meet the requirement of the rated power when the equipment is working normally. The result proves the effectiveness of the control system of the phase control rectifier, and has a good value in engineering application and popularization.

Keywords: Phase controlled rectifier, DSP, PID closed loop regulation, DC voltage detection, temperature detection, phase shift control

1. Introduction

Due to the continuous development of China's coal industry, the continuous increase of coal production, the continuous application of high-power coal mining units and transport equipment, coal mine power supply is becoming more and more electrified. Power electronic devices such as converters have been widely used in all aspects of mine [1]. The front part of these power electronic devices basically use uncontrolled rectifier combined with charging contactor, and then use IGBT inverter to drive AC motor. The power load of power supply of coal mine has great change, frequent start and stop, and even temporary stop, which makes the power grid fluctuate relatively large. Voltage fluctuations often cause great impact on mine production, ventilation and transportation, and are very harmful to the use of electrical equipment. Sometimes, it will damage equipment and cause accidents. This requires that the phase controlled rectifier of mine has a very reliable voltage regulation performance, enabling the electrical equipment to work at a rated voltage under complex working conditions and ensuring the safe and reliable operation of electrical equipment [2]. With the continuous increase in the length of fully mechanized coal mining face and mining area, the production of coal mine is increasing. The power (capacity) of the equipment in the mining face is increasing. The transmission distance is constantly lengthening. The voltage loss of the cable line is getting bigger and bigger. The demand for the output voltage of the rectifier is getting higher and higher. This requires that the DC side output voltage of the rectifier vary according to the mining depth of the coal mine.At present, the underground power supply line system adopts the highest 10kV voltage in the country. The actual input to the mining face voltage is generally divided into 2.3kV, 3.3kV, 4.16kV, 5kV and 6kV several [3]. The power supply voltage of the

coal mine requires that the output voltage of the rectifier can be adjusted.
Nowadays, the PWM rectifier control system in the market is immature in terms of high power and high voltage technology, low energy efficiency, and has the disadvantages of inconvenient installation and debugging[4]. The demand for explosion of electrical equipment is very high in coal mine, so the cost of installing flameproof input is very high. Therefore, it is necessary to develop a new type of coal mine phase controlled rectifier control system with low cost, reliable voltage stability and good phase control performance.

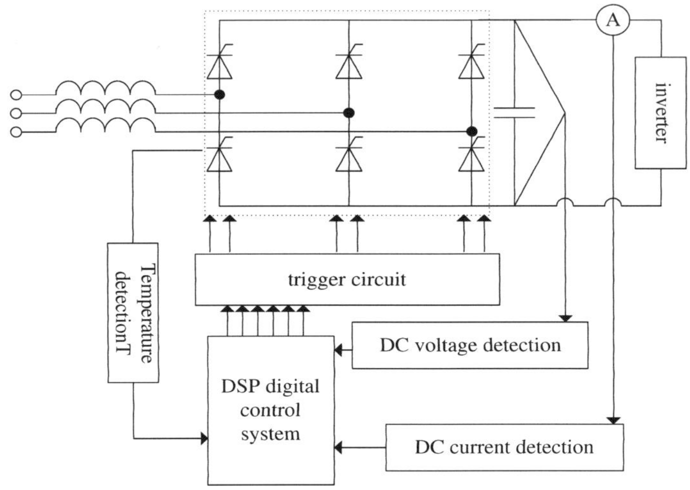

Fig.1.Phase controlled rectifier circuit

2. System composition and hardware circuit design

2.1.System composition

The three-phase AC input power supply is supplied to the three-phase inverter for the mining equipment through the stable DC voltage output of the phase controlled rectifier. With the increasing level of the three-phase AC input voltage, the phase controlled rectifier can automatically adjust the stable DC voltage required by the output inverter or the load. When the power load of the mine is changed greatly, the phase control rectifier can stabilize the DC voltage when the power grid fluctuates[5-8].
The designed phase control rectifier system is shown in Fig.1. It is mainly composed of 3 parts: (1) main electric route three-phase bridge full control rectifier circuit and inverter; (2) DSP digital control system; (3) various detection circuits: grid side current detection circuit, grid side voltage and DC voltage detection circuit, temperature detection circuit.

2.2.DSP digital control system

The block diagram of the DSP digital control system is shown in Figure.2. The core chip uses TMS320F28035 fixed-point DSP. By using F28035's rich peripheral interface and 12 bit high precision AD, we can accomplish all kinds of sampling signal, trigger pulse and protection state signal input in mine phase control rectifier drive control system. The DSP digital control system is the core part of the whole mining phase controlled rectifier control system [9-10]. It is mainly composed of 4 parts: DSP digital system, analog input circuit, man-machine interface circuit and thyristor drive circuit, thyristor temperature detection circuit, fault detection circuit and so on.

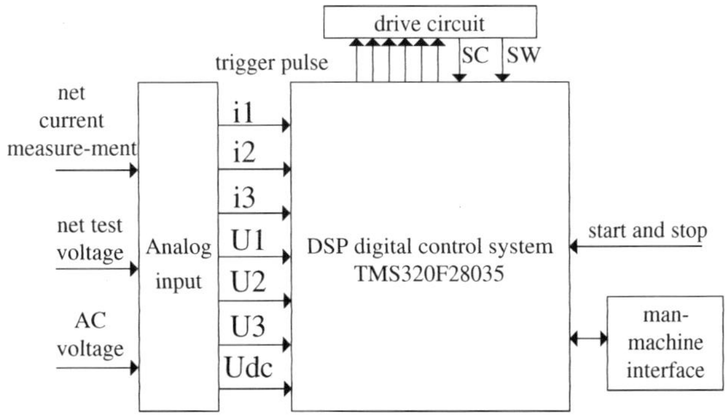

Fig.2. Block diagram of DSP control system

(1) DSP digital system: it is composed of TMS320F28035, digital input signal of external terminal and so on. It can perform all kinds of functions, such as digital calculation, external terminal control, starting and stopping control and operation of the mine phase controlled rectifier circuit. (2) Analog input circuit: it mainly consists of network side voltage, current and DC voltage detection and conditioning circuit. The main circuit of voltage and current detection is to convert the strong signal to the weak signal for digital sampling by DSP. (3) Thyristor drive circuit: converting the trigger pulse signal from the DSP system into a pulse signal that can drive the thyristor. When the thyristor is too high or overcurrent, the fault signal is fed back to the DSP control system. (4) Man-machine interface circuit: It is mainly composed of CAN communication, upper PC display and operation keyboard. A variety of given instructions and parameters can be set by the keyboard, such as the starting and stopping of the phase controlled rectifying circuit of the mine, the given DC output voltage value and so on. It can display the various states of the phase controlled rectifying circuit of the mine and record the history of the phase control rectifying circuit of the mine.

2.3.Hardware design of various detection circuits

2.3.1.Network side current detection circuit.

The detection of network side current is very important for mine phased rectifier control system. Its value will be used as the instruction current of synchronous PI decoupling current control. The reliability and accuracy of detection is the first consideration. The sensor detection scheme is selected here. In a three-phase balance system, the line voltage is 380 V, and the maximum per phase current is not more than 400 A. Considering some margin, the selected current sensor is a three phase AC current sensor of LAC300-S model of LEM company. The initial stage of the sensor is insulated. The conversion rate is 1:4000. Measurement range 400A. It is used for three phase precision measurement of DC, AC and pulse current. The value of the rated output current is 100mA, and the voltage of the power supply is + 15V.

As shown in Fig.3, the left side of the graph is the sensor. The three phase lines of the power supply are passed through the three measurement holes respectively. The M1, M2 and M3 in the sensor are three phase measurement current output respectively. A phase current conditioning circuit is given in the diagram. Because of the three-phase balance system, the three-phase current regulating circuit is consistent. The detection values of each phase current pass through the magnetic beads (restraining high frequency interference and peak) and flow through the sampling resistance. The current signal becomes a voltage signal. Then through the TL082 operation and amplifier conditioning, the output voltage is between 0-3.3V. Finally, the output voltage is input to the AD port of DSP after the DAN217 clamping protection.

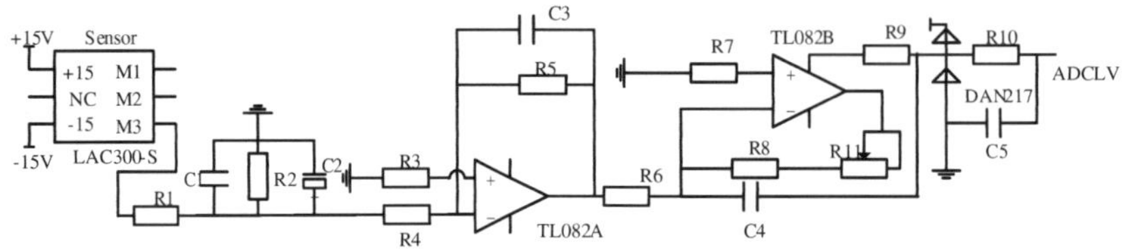

Fig.3. Network side current detection circuit

2.3.2.Network side voltage and DC voltage detection circuit.

The voltage value of the net side is used as the instruction of the synchronous PI decoupling current control, and the decoupling algorithm is realized. The line voltage rating of the test network side voltage is 380V. Therefore, the type of voltage sensor is TBV10/25 A. The primary and secondary of this type of sensor is insulated.It can be used to measure DC, AC and pulse voltage. The rated input current is 10 mA, the output current is 25mA, and the supply voltage is 15 V. The three-phase voltage detection circuit on the net side is consistent. As shown in Fig.4, a circuit diagram for detecting the voltage of the two line voltage of the net side voltage is given. The two line voltage is connected by the input side current resistance R1 of the sensor. After the sampling resistance R2, the line voltage becomes the voltage signal. After the operation of capacitor filter and TL082 operation, the voltage signal is output to the AD port of DSP[10-15].

The value of the DC voltage is used as the external loop control feedback instruction controlled by the synchronous PI decoupling current. The function of the DC voltage which is stable in the output of the phase control rectifier is realized. The detection circuit is the same as the network side voltage detection circuit. Only the value of the input current limiting resistance R1 and the sampling resistance R2 are different.

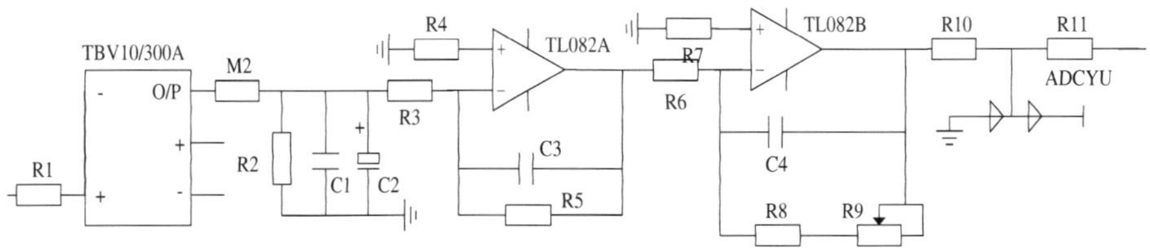

Fig.4. Network side voltage and DC voltage detection circuit

2.3.3.Temperature detection circuit.

The driving control system of the mine phase controlled rectifier is installed in the closed electric control box of the shearer, which itself is natural cooling. The heat dissipation effect is strengthened by its tight installation on the outer wall of the water cooling system of the shearer. But at work, the temperature will rise sharply. Even if the radiator is strengthened by the shearer, the cooling effect will not be ideal if it is used for a long time. It will make the temperature too high and damage the drive control system. So it is very important to detect the temperature of the thyristor. The PT100 temperature sensor is used to detect the thyristor. As shown in Fig.5, the PT100 is in full contact with the surface of the system radiator. When the temperature of the driving control system of the phase control rectifier of the mine is changed, the resistance of PT100 also changes. The change of the resistance value can be measured by the circuit shown in Fig.5, so that the temperature of the drive control system is measured. In Fig.5, TL431 constitutes a voltage stabilizing circuit, providing a precise +5V voltage. The operational amplifier TL082 and the resistor are composed of a 200mA constant current source circuit. A constant current source is used to convert the PT100's resistance to a voltage signal. After using AD620 to adjust the signal, the signal is sent to the AD port of DSP for conversion.The temperature detection circuit will feed back the temperature of IGBT in real time, and the real-time monitoring and temperature protection will be realized by using DSP.

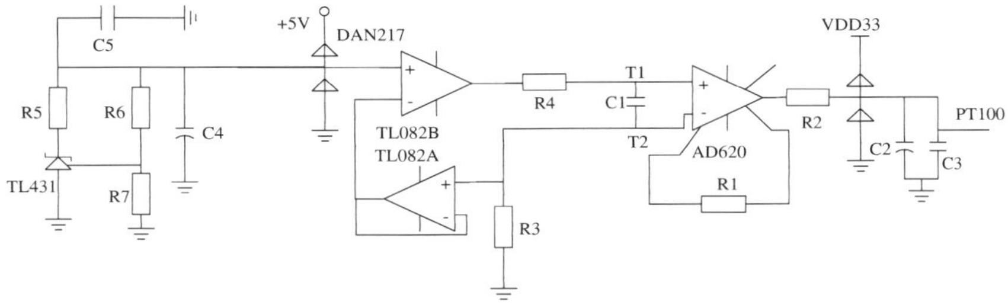

Fig.5. PT100 temperature detection circuit

3. Software design

The software program of drive control system is mainly composed of main program, AD timer interrupt service routine, capture interrupt service program, CAN communication interruption program, waveform generator interrupt program and so on. The software flow chart is shown in Fig.6.

The main program mainly completes initialization variables and data settings, initialization settings of special function registers and external event management registers, interruption setting, interruption and other functions, and completes the signal sampling of external terminals. It manages the starting and stopping of the phase control converter, the reset signal of CPU, the output fault signal and so on.

The CAN interrupt service program mainly completes the communication with the operation keyboard, and sends the running state and fault information of the drive control system of the mine phase control rectifier to the display. Receive the keyboard setup command at the same time. Capture interrupt service program, which mainly completes the sampling and calculation of the phase, phase sequence and frequency of the power grid. Waveform generator interrupt service program, which mainly completes the update of the cycle register and duty ratio, the calculation of duty ratio and the generation and output of the SVPWM wave.

AD timer interrupt mainly complete the following work: (1) Sampling calculation of three phase AC current, sampling calculation of three phase AC voltage, sampling calculation of DC voltage, sampling calculation of IGBT temperature. (2) The calculation of dead zone compensation, the calculation of the cycle register value of the waveform generator, the calculation of the value of the phase accumulating, and the calculation of overload over current.(3) The calculation of the synchronous PI decoupling current control algorithm.

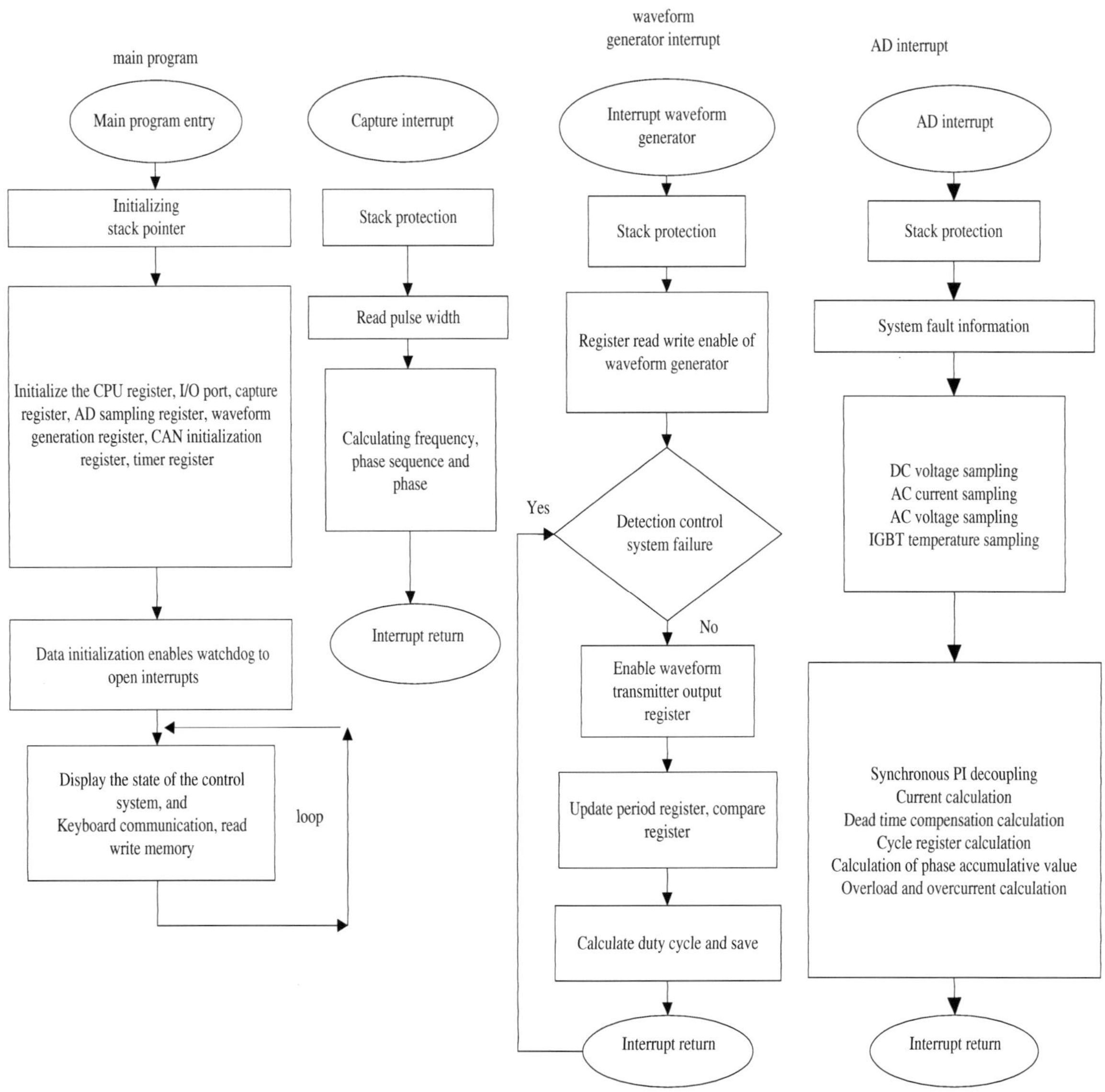

Fig.6. Software program for driving control system

4. Experimental research

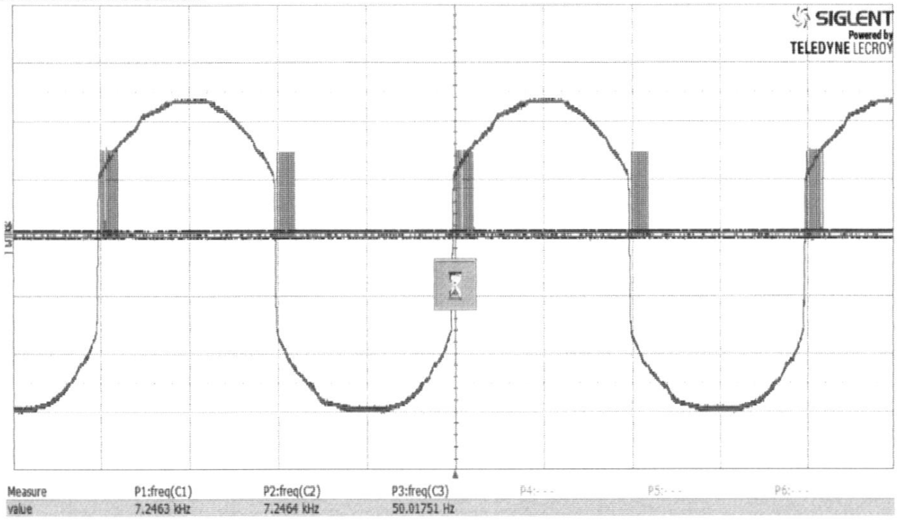

Fig.7. The input waveform of the synchronous signal and the output waveform of the trigger pulse

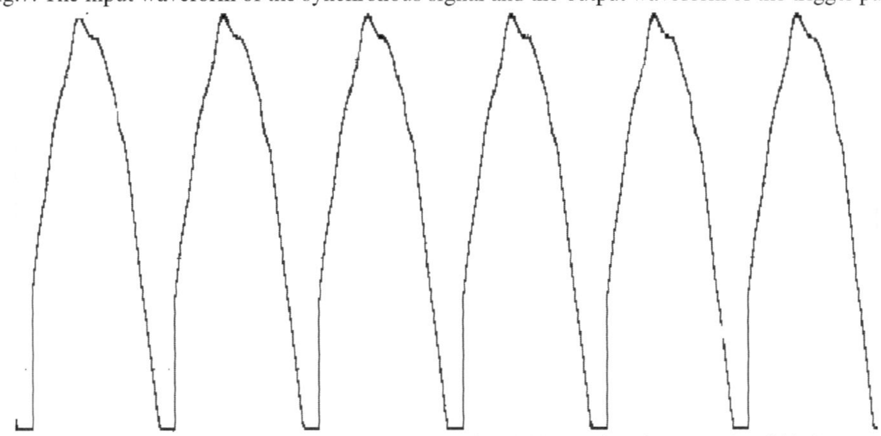

Fig.8. Output waveform with an input voltage of AC380V and a trigger angle of 30 degrees

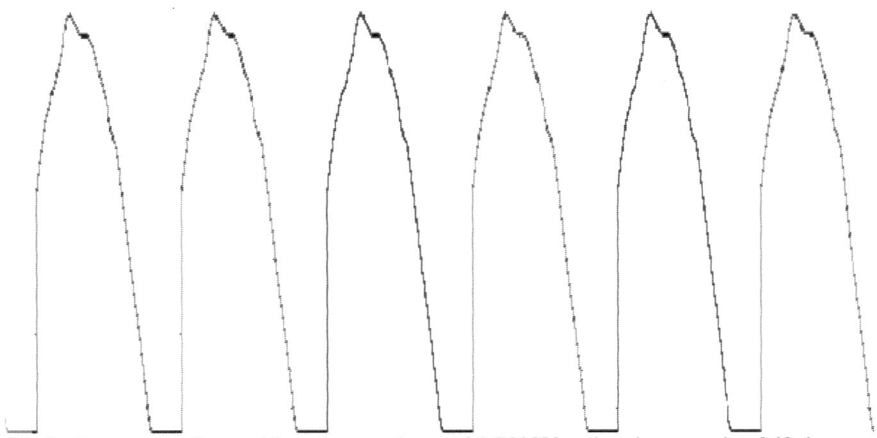

Fig.9. Output waveform with an input voltage of AC380V and a trigger angle of 60 degrees

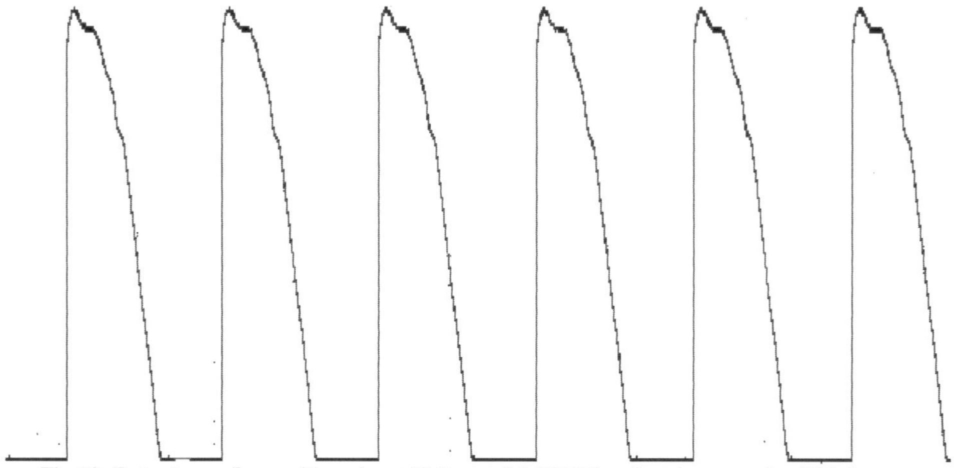

Fig.10. Output waveform with an input Voltage of AC380V and a trigger angle of 90 degrees

In order to verify the reliability of the control system of the mine phase controlled rectifier, the driving control system of the phase controlled rectifier is studied and applied to the underground coal mine. The main parameters of the system are: the input three-phase AC voltage of the main circuit is AC380V, the synchronous voltage of the synchronous transformer is AC20V, and Fig.7. is the waveform diagram of synchronous input voltage signal and trigger pulse sequence.

From the waveform diagram, we can see that the driving control circuit can output reliable trigger pulses, which can effectively drive thyristor and output stable DC voltage. Fig.8-10, respectively, are the output voltage waveform of the DC side when the trigger angle are 30 degrees, 60 degrees and 90 degrees without adding filter capacitor. From the waveform chart, we can see that the phase controlled rectifier can control the output voltage well, and output the stable DC voltage needed.

5. Conclusion

In view of the characteristics of mine power supply and electrical equipment, a full digital intelligent control phase control rectifier control system is designed. The system can not only output the DC voltage of any desired setting, but also stabilize the output DC voltage, and it can be applied to the high input voltage level of the mine to adapt to the larger input power and larger load. It can meet the increasing input voltage of mining equipment with the increasing mining depth of coal mine. It has the characteristics of simple circuit, strong modularization, flexible parameter setting and good dynamic performance. It has improved the control performance, adaptability, flameproof performance and reliability of the operation of mine equipment, and has a good value in engineering application and popularization.

Fund Project: Hebei provincial science and Technology Department (16274603,16214408), China Coal Industry Association (MTKJ2017-309)

6. References

[1] Wang Zhanquan.Current Situation and Development of Power Supply System in Coal-mine underground Face in China *J. Shanxi Coking Coal Science and Technology*, 2005,5,3.

[2] WU Wen-ming, Zhang Sheng-gang, Wang Bing.A Brief Discussion on the Existing Problems and Solutions of Coal Mine Power Supply System *J. Industry and Mine Automation*, 2009,9,10 (in Chinese).

[3] Yan Dong, Liu- Shigang, Fu Jianwei. Analysis of fluctuations in supply voltage of mines and its inhibition *J. Shandong Coal Science and Technology*, 2009,4,15.

[4] Li Yuchao, Gao- Qinxiang.Design and Development of Three-phase Full- control Bridge Rectification Experiment Device *J.Modem Electronics,* 2006,10,6 (in Chinese).

[5] Deng Yonghong. Design of drive and control system of 4-quadrant converter for traction sheater *J. Industry and Mine Automation*, 2016,6.

[6] Ren Zhongxiao. With imported KJ009 instead of imported components *J. Equipment Management and Maintenance*, 1992,3,1.

[7] Wang Zhaoan, Huang Jun. Power Electronics *M. Machinery Industry Press*, 2009: 56-58.

[8] General family .KJ042 pulse train modulation generator *J.Electrical Times*, 1992,11,16.

[9] Zhao Liyong, Zhang Quanzhu, Deng Yonghong. Research and design of phase control circuit for auxiliary converter system of electric locomotive *J. Journal of North China Institute of Science and Technology*, 2012,1,15.
[10] Zhao Liyong, Zhang- Quanzhu, Deng Yonghong.Research and Design of Phase Controlled Circuit for Auxiliary Converter of Electric Locomotive *J.Journal of North China Institute of Technology*, 2012,1,15.
[11] Li Changshun.Design of Thyristor Trigger Circuit Based on DSP *J.Laboratory Science* ,2010,4,28.
[12] Lv Baoliang. Design of Three-phase Voltage-source PWM Rectifier Based on DSP *D. Harbin Institute of Technology*, 2007,3,1.
[13] Yang Junlian.Design of Three-Phase PWM Rectifier PI Regulator Based on DSP *J.Communications Power Technology*, 2012,3,25.
[14] Wu Xiaochun, Tian Jiu-ting, Wang Zhen.Dual-closed-loop control system simulation of three-phase voltage source PWM rectifier *J.Journal of Lanzhou Jiaotong University*, 2017,6,15.
[15] Shao Zelong, Shi Mingquan, Han Zhenfeng.Design of PWM Rectifier Based on Fuzzy PI Control *J. Power Technology*, 2016,1,20.

Research of Intelligent Lead-acid Batteries Charger in Coal Mine

Yonghong Deng[a,b], Zhishan Liang[a], Haiyu Chen[b], Yuanhong Li[b] and Quanzhu Zhang[b]
[a]College of Geophysics and information Engineering,China University of Petroleum(Beijing),Beijing 102249,China
[b] School of electronic and information engineering, North China Institute of Science and Technology， East Yanjiao, Beijing, 101601, China

dyhsyjdyx@163.com; 1195078260@qq.com; qzhzhang@ncist.edu.cn

Abstract. The scheme of intelligent lead-acid battery charger is studied, and the advantages and disadvantages of the 5 kinds of intelligent charger are compared and analyzed. A most intelligent and fast charger scheme is put forward. The intelligent charger has the advantages of bidirectional energy flow, small volume, light weight, convenient voltage regulation, small starting current, fast dynamic response, high intelligence, wide input voltage and fast charging. The intelligent charger is made up of various power conversion units. PWM technology is adopted to transform the input three-phase AC380V/AC660V to the DC0V to DC550V through voltage transformation. The intelligent charger scheme is novel, advanced in technology and superior in performance. It can be popularized and applied in the coal mine industry.
Keywords: Mining electric locomotive,lead-acid batteries,intelligent charger,high-frequency Transformer,phase-shift PWM technique,charging modes

1. Introduction

From coal mine safety production point of view, mining electric locomotive is an important mine mechanical and electrical transport equipment. Lead-acid battery electric vehicles have mobility, flexibility, low power supply costs, no sparks and other advantages of explosion-proof. Therefore, the majority of rail electric locomotives used lead-acid batteries[1-5]. Lead-acid battery is relying on the charging device to charge it repeatedly, to achieve the re-use of electrical energy. Therefore, the charger device is an important coal mine used with the lead-acid battery pack. The performance of the charger device directly affects the operational efficiency, service life and reliability of the lead-acid battery. The design of the charger for mining electric locomotives is of utmost importance. This equipment has drawn more and more attention from coal mining enterprises in recent years[6-10].
According to the fast charge theory, the charging process releases the polarization voltage of the battery pack to the power grid instantaneously through the charger, or the battery pack is normally discharged back to the power grid through the charger instantaneously. This charging method increases the charging current or can maintain long-term charge current. Overall, this charging method shortens the charging time [11-12]. Both of the above chargers do not discharge the battery rapidly and are also difficult to charge all the battery packs. According to the actual requirements of mine chargers and the need for fast charging, it is of great significance and broad market prospects to study the mining of lead-acid battery packs for high-frequency smart chargers with bidirectional flow of energy for fast charge and discharge.

In summary, mine lead-acid battery smart charger should have the following characteristics: (1) Taking the actual
situation of coal-mine lead-acid batteries in series parallel into account, the charger requirements of lead-acid battery pack, it is necessary to charge all battery packs. The charging voltage output by smart charger should be continuously adjustable between DC0V ~ DC550V. (3) Taking the coal mine power supply and underground harsh environment into account, the charger should automatically adapt to AC380V / AC660 two kinds of input voltage. (4) Taking the requirements of lead-acid battery charging mode into account, the charger can realize stable voltage and can realize multi constant current charging function.(5) The charger should meet the requirements of the charging of the lead-acid battery group in coal mine. The equipment is small in volume, light in quality, easy to move, transport and repair.(6) Considering the explosion protection requirements of the underground, the charger is natural heat dissipation. (7) Charger has the ability to automatically adapt to underground power supply voltage fluctuations.
Therefore taking into account the advanced technology and engineering practicality, the study of rapid charge and discharge of lead-acid batteries for mining high-speed smart charger, has important significance and broad market prospects.

2. Comparison of several schemes for intelligent charger control circuit

2.1.An intelligent charger scheme for input power frequency transformer and thyristor phase control
This intelligent charger scheme adds a thyristor phase controlled voltage regulator charger to the input power frequency transformer, as shown in Fig.1, which is widely used in the industry at present. It is a typical AC - direct (AC/DC) voltage conversion mode. The voltage regulating part is realized by phase controlled rectifier with thyristor. The electrical isolation is realized by the power frequency transformer. The primary winding of the transformer by different Y- delta connection, to meet the requirements of AC380V/AC660 two power supply.
The advantages of the power supply of the circuit is relatively simple, transform more directly, not only to achieve the electrical isolation of input and output, but also achieved by the AC voltage to DC voltage output transformation, while increasing the control of output voltage and current regulation, have a certain effect.
The shortcoming is also obvious, because the output voltage of thyristor phase controlled rectifier is 300 Hz, and the pulsating DC voltage is larger, so the charging of battery pack still has some unfavorable factors. If a smooth filter reactor is added to the DC side, the volume and weight of the charger will be increased. In this scheme, the volume and weight of the equipment are greatly increased because of the existence of the input transformer, and the quality of the added flameproof shell is heavy (more than 750 kg). Therefore, the rough charging mode, large volume and heavy weight are the disadvantages of the traditional battery charger, which need to be improved.

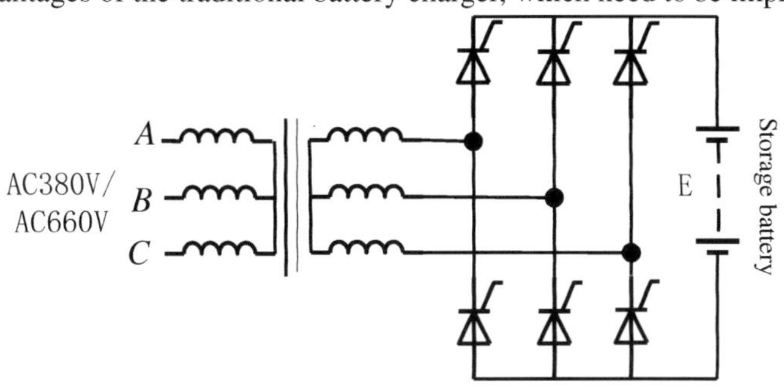

Fig.1. Electrical principle block diagram of intelligent charger for thyristor phase controlled rectifier

Secondly, the multi section constant current charging process of the battery group can not be realized. When the battery group is badly in loss of electricity(Excessive electrical time leads to low output voltage) ,the charging current must be strictly controlled, and the charging current is limited by reducing the output voltage of the charger, and the pulse of the charging voltage should be as small as possible. Otherwise, the charging current is too large, the battery is overheated and the plate is aging. The life of the battery is greatly shortened. The traditional charger basically uses analog electronic control circuit, it is difficult to achieve multi-stage constant current charging function, and it is impossible to achieve charging current and battery voltage adaptive charging condition. This is also a significant drawback of traditional charger.

Finally, the two input voltages of the AC380V/AC660 can not be automatically adapted to the two types of input voltages. It is adapted to the connection mode of changing the power supply to the original side winding of the transformer.

2.2.Input power frequency transformer with uncontrolled rectifier and IGBT chopper voltage regulation intelligent charger scheme

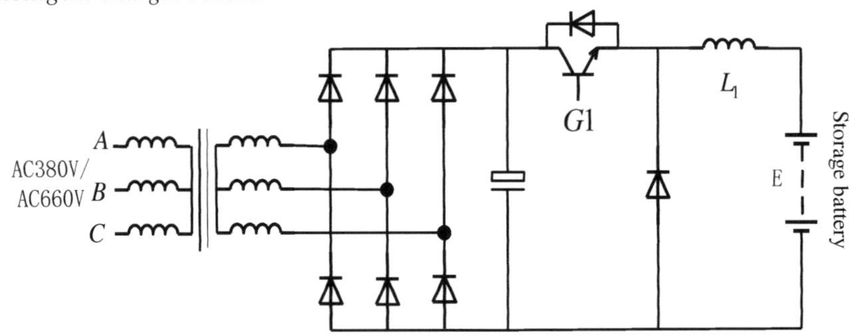

Fig.2. Electrical schematic diagram of IGBT chopper intelligent charger

The intelligent charger scheme is called input power frequency transformer plus uncontrolled rectifier and IGBT chopper voltage regulator intelligent charger scheme as shown in Fig.2. The scheme is a "AC direct to direct (AC/DC/DC)" charging scheme. The chopped wave step-down (Buck) converter will cut the DC voltage of the power frequency rectifier to less than DC550V (it can be continuously regulated according to the charging voltage, such as DC0V ~ DC550V), which is used to charge the lead-acid battery group. The voltage regulation part is realized by the IGBT chopping mode, and the electrical isolation is realized by the power frequency transformer. The primary winding of power transformer uses Y- delta different connection, to adapt to the requirements of AC380V/AC660V two power supply.

The advantage of this intelligent charger is that it relies on the power electronic power converter to complete the voltage transformation. This scheme achieves a constant voltage and constant current charging multi segment. The voltage is continuously adjustable, the technology is advanced, and the degree of automation is high. But the disadvantage is that the electrical isolation is realized by the power frequency transformer, and the inductance is added to the DC side, so the charger is large in volume and heavy in weight. The scheme, like the first one, can not automatically adapt to the two input voltages of the AC380V/AC660V, which is accomplished by changing the connection mode.

2.3.Uncontrolled rectifier and chopper plus H bridge IGBT inverter and high frequency transformer isolation and rectification intelligent charger scheme

The scheme of intelligent charger called uncontrolled rectifier plus IGBT chopper plus H bridge inverter and finally added high frequency transformer and rectifying intelligent charger scheme is shown in Fig.3. The scheme is a "cross - Direct - to - Direct - to - direct (AC/DC/DC-DC/AC/DC)" charging scheme, consisting of two sets of converters of AC/DC/DC and DC/AC/DC. Automatically adapt two power supply AC380V/AC660V level AC/DC/DC converter, the DC voltage of the chopper controlled rectifier will not reduce, it realizes the input AC380V/50Hz (power frequency) or AC660V/50Hz (frequency) AC voltage into a DC voltage to the DC voltage of the converter. The secondary converter will rectify the intermediate DC voltage inverter and output the charge voltage required by the storage unit. The voltage regulating part is mainly realized by chopper and high frequency transformer, and the isolation of electrical apparatus is realized by high frequency transformer.

This scheme can realize constant voltage and constant current charging multi segment. The voltage is continuously adjustable, the technology is advanced and the automation level is high. It can automatically adapt to two input voltages of AC380V/AC660V. The adoption of high frequency transformer reduces the volume of the device and lightens the quality. It integrates voltage regulation and electrical isolation into one, and the integration degree is high.
However, the intelligent charger scheme can not achieve both the two-way flow of energy and the discharge of the lead-acid battery. The charging speed can be further improved.

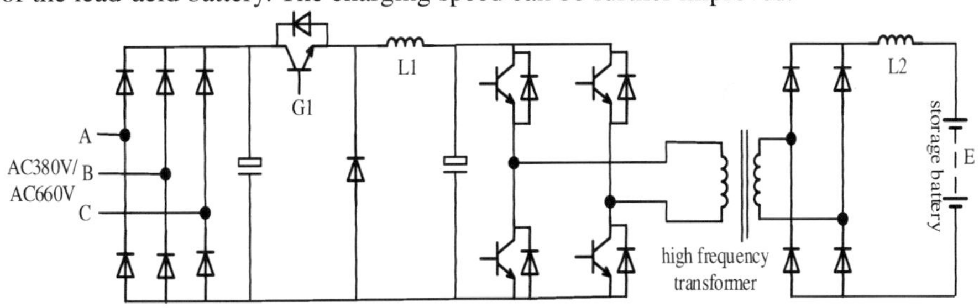

Fig. 3. IGBT chopper and H bridge inverter with high frequency transformer intelligent charger electrical principle diagram

2.4. Uncontrolled rectifier and H bridge IGBT inverter and high frequency transformer isolation and rectification intelligent charger scheme

This scheme is shown in Fig.4 for the uncontrolled rectifier plus the H bridge IGBT inverter and the high frequency transformer isolation and rectification intelligent charger scheme. The scheme is a "AC direct to AC (AC/DC/AC/DC)" charging scheme. The H bridge converter transfers the DC voltage to DC300V after power frequency rectification, which can be regulated continuously according to the output voltage, such as DC0V to DC300V, to charge the lead-acid battery group. The voltage regulation part is realized by H bridge inverter, and the electrical isolation is realized by high frequency transformer.
The advantage of this charger is that it reduces the volume of the device and reduces the quality. It integrates voltage regulation, electrical isolation and integration, and relies on power electronics technology and microcomputer control technology. Constant current and constant pressure charging is convenient. It is advanced in technology and high in automation. It automatically adapts to two kinds of input voltage of AC380V/AC660V.
The same kind of smart charger will be more expensive than the traditional charger. In particular, to adapt to two kinds of power supply (AC380V/AC660V), the charger has to output the DC voltage of DC0V to DC300V. The variable ratio of the high frequency transformer is designed according to the minimum value of the power supply (The value of the AC380V's lower limit fluctuates to 75% of the voltage value of AC285V). At the same time, according to the maximum output DC voltage DC300V, the calculation change ratio is 1.3:1; at this time, the original side winding current of the high frequency transformer is 92A according to the rated output current 120A. The original winding and the secondary winding current are basically the same, increasing the difficulty of the design of the high frequency transformer, including the increase of weight and volume.
At the same time, it is assumed that the input power is AC660V used to charge the battery group of the coal mine DC48V. If the control system uses full bridge phase shifted PWM control technology, the phase shift angle of the control system is less than 10 degrees. To meet the charging voltage requirements of all the battery pack, the phase shift range of the charger control system is at least from 10 to 180 degrees.Its stability efficiency remains to be verified.

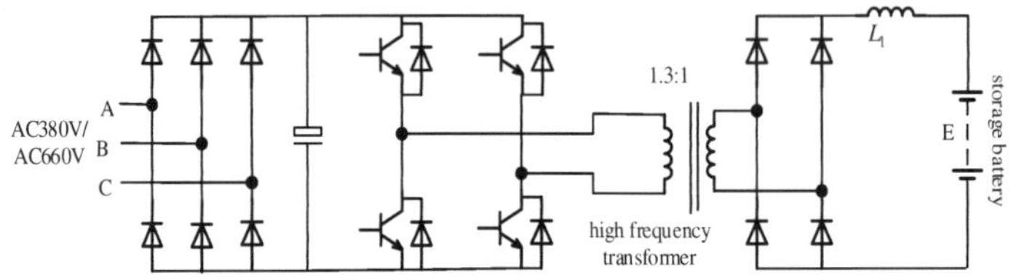

Fig.4. H bridge IGBT inverter with high frequency transformer intelligent charger electrical principle diagram

2.5.Intelligent fast charger scheme for multi power unit

This intelligent charger scheme is actually an improvement in the above charger scheme. The entire main back route is composed of multi power units, as shown in Fig.5. It is mainly composed of 6 parts: PWM rectifier, bidirectional DC/DC converter, bidirectional H bridge inverter 1 (fast rectifier module 1), high frequency transformer, fast rectifier module 2 (bidirectional H bridge inverter 2), and control systems. The variable ratio of the high frequency transformer is set to 1:1.

When the system is in the charging operation condition, energy flows from the power grid to the battery pack:Power grid→PWM rectifier→Bi-directional DC/DC converter→H bridge inverter 1→High frequency transformer→Fast rectifying module 2→Battery pack; When the system is in the discharge operating condition, the energy flows from the battery pack to the grid:Battery pack→Bidirectional H bridge inverter 2→High frequency transformer→Fast rectifying module 1→Bi-directional DC/DC converter→PWM rectifier→Power grid.

The PWM rectifier[13], when charged, works in the condition of rectifying, and can output a stable DC voltage, which can be adapted to the input power with a large fluctuation range. When working in the inverter, the energy of battery pack can be quickly fed back to the power grid without external discharge resistance unit, which improves the control performance, flameproof performance and reliability of the charger. The input current is sinusoidal and the unit power factor of the system is improved.

To sum up, the charger used in this scheme is automatically compatible with two kinds of input voltage, and the output charging voltage DC0~DC550V and charging current DC0A-DC120A are continuously adjustable, which can charge all specifications of battery pack. By calculation, the duty cycle of Buck chopper can be fixed, the phase shifting angle of 1 H bridge inverter can be adjusted automatically, the duty ratio of coordinated control buck chopper and the phase shift angle of H bridge inverter 1 will be charged to the battery pack. This strategy is easy to implement, control simple and dynamic response fast. All power devices work at the best state and consume less, making the charger efficient and reliable, enabling fast charging and discharging, energy saving and environmental protection.

Through the comparison of the above 5 kinds of intelligent charger schemes, combined with the current technology level and practical application, we think that fifth kinds of intelligent charger schemes are the preferred charger schemes in the current lead-acid battery charger.

3. Conclusion

The intelligent charger scheme for lead-acid battery group is proposed in this paper. It integrates power electronic converter technology, microcomputer control technology, field bus network technology and mechatronics technology. The problem of over voltage and over current charging, resulting in battery heating and scrapping and slow charging speed in the charging process of coal mine lead-acid battery group is well solved. Through the analysis and experiment, the following conclusions are drawn.

(1)The smart charger of the lead-acid battery group is small in volume and light in weight, which realizes the miniaturization and lightening of the intelligent charger of the coal mine lead-acid battery group. The same output power (capacity), compared with the traditional charger, can be reduced to the original 1/3, and the weight can be reduced to the original 1/4.

(2)In the control system, the full digital microcomputer control and phase shift PWM control technology are used.

IOP Conf. Series: Journal of Physics: Conf. Series **1072** (2018) 012009 doi:10.1088/1742-6596/1072/1/012009

(3)In the improvement of the main circuit, the bidirectional flow of energy is realized, and the speed of charge and discharge is fast.

(4)In the aspect of function expansion, intelligent charging is realized by microcomputer control. The charging characteristics are multi segment and adaptive, and the charging curve can be adjusted by multivariable control, and the service life of the battery is prolonged.

(5)It can automatically adapt to the input of two kinds of power supply. The charging voltage can be adjusted continuously, and the charging current can be set according to the need.

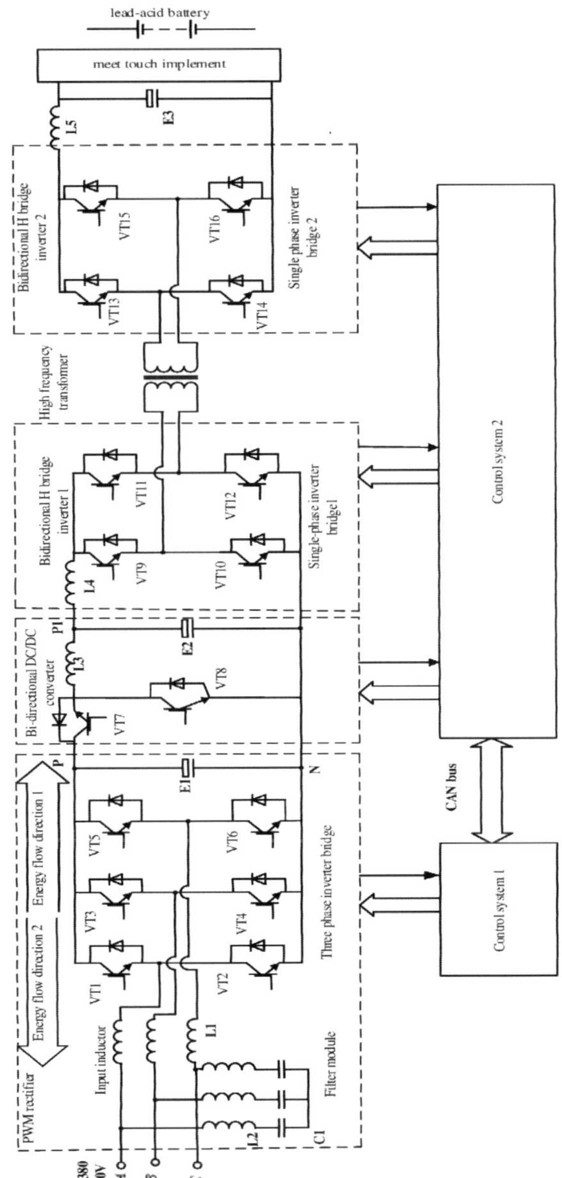

Fig.5.Electrical schematic of a multi power unit intelligent charger

Fund Project: Hebei provincial science and Technology Department (1627460316214408), China Coal Industry Association (MTKJ2017-309)

4. References

[1] Deng Yonghong, Pan Yumin, Zhang Quanzhu, Huang Chengyu. Mine lead-acid battery pack smart charger J. Power Technology, 2013,37(10):1801-1803.

[2] Deng Yonghong, Xu Mingyan, Li Xuezhe. Research on high frequency intelligent charger of lead-acid battery in mine J. Coal Science and Technology, 2015,47(3): 132-135.

[3] Zhang Quanzhu, etc. Mine electric locomotive lead-acid battery smart charger J. Case Automation, 2009;10(28): 121-123.

[4] Zhang Quanzhu, Zhang Feiping, Pei Cui-ling, Deng Yonghong. A new type of mine lead-acid battery smart charger J. Mining Industry, 2013;39(3): 35-39.

[5] Zhang Xiaoqun, Gao Yanxia, etc. Mine motor vehicle battery smart charger development [J]. Coal Mine Machinery, 2007;1(28): 121-123.

[6] Han Xiangfeng, Li Zhen wall, etc. Mine lead-acid battery high-power smart charger development J. Coal Mine Machinery, 2009;1(30): 132-134.

[7] Li Ning, Zhang Hongtao. Mine special type lead-acid battery charging technology J.Coal Mine Machinery, 2013;34(5): 250-252.

[8] Liu Yingcan. Mine electric locomotive battery system using the status quo analysis and upgrading of the direction of research J. Coal Engineering, 2012;10: 121-122.

[9] Guo Fengyi, Miao Chuanhai, Zhang Jihua.Design and implementation of intelligent lead-acid battery charger J. Journal of Computer Applications, 2012;21(3): 76-79.

[10] Feng Chuan-hua.Study on mine embedded intelligent charger based on fuzzy control J. Journal of Changchun Institute of Technology, Natural Science Edition, 2011;12 (1): 58-62.

[11] Zhou Zhimin, Zhou Jihai, Ji Aihua. Charger circuit design and application M.Beijing: People's Posts and Telecommunications Press, 2005.

[12] Wang Gonggong. Lead battery operation and maintenance M.Beijing: Water Conservancy and Electric Power Press, 2006.

[13] Chen Yao, Tong Yibin, Jin Xinmin. New algorithm for harmonic analysis of SVPWM based on PWM rectifier J. Proceeding of the CSEE, 2007,27(13): 76-80.

Discussion on the Over-frequency Generator Tripping Scheme of the Power Grid

Yiting Guo[1], Haipeng Nan [2], Xin Guan [3], Luochang Wu [4]

No. 5 Jinhua South Road, Beilin District, Xi'an City, Shaanxi Province,China

E-mail: 376423022@qq.com

Abstract. The over-frequency generator-tripping and under-frequency load shedding devices play an important role as the third defensive line for frequency emergency control measures, which have a great significance for maintaining the power grid to operate safely and stably. At present, the abnormal operation frequency of hydropower generating units is mainly set based on the standard of the thermal power unit. Thus when the proportion of hydropower is too large, it can easily cause over-frequency generator tripping and under-frequency load shedding devices act one after the other, which may causes the frequency collapse. In this paper, an over-frequency cutting scheme for power grid with hydropower unit is formulated. The paper analyses and summarizes the over-frequency protection of the hydropower unit, by studying the 10 machine 39 node model with PSD-BPA software. The paper also gives a reasonable frequency action value of the high frequency cutting machine. The results show that the first-round operation value of the high-frequency cutting device of the hydropower unit should be higher than the operation value of the thermal power unit. With the first-round operation value of the high-frequency cutting device set near 52.0Hz, it can effectively avoid both of the over-frequency generator tripping and the under-frequency load shedding device, and also ensure the safety and stability of the power grid. In the hydro-thermal power grid, because of the frequency of hydropower unit cuts increasing, the power grid should remove the thermal power unit firstly, which is more beneficial for the power system frequency to stabilize rapidly.

1. Introduction

China is rich in hydropower resources and hydropower has become an important part of China's power grid. In some regions, hydropower units account for a large part of the power grid, such as the power grid in Xishuangbanna, Guizhou, Sichuan Jiushiya regional power grid [1~2]. The power grids in hydropower-rich regions mostly are a kind of sending system and far away from the load centre. Because of the large capacity of hydropower units, when the fault of the power supply network in the area which hydropower accounts for a large part of the power supply, it may lead to large frequency fluctuations[3].

If the over-frequency generator tripping and under-frequency load shedding (UFLS) devices in the power grid are not reasonable, the large unit will trip and be over-cut when the frequency of the sending system exceeds the allowable operating range. Also, it may result in an insufficient regional output which can make the system frequency to drop sharply and trigger the UFLS device. Then the frequency of system may rise significantly that triggers the operation of the over-frequency generator tripping device. This cycle process will lead to the collapse and blackout for the power grid, which brings great economic losses [4~7]. Therefore, the operation value of over-frequency generator tripping should be set reasonably to avoid the occurrence of large-scale power outages caused by repeated operation of over-frequency generator tripping and UFLS devices, so as to ensure the safe and stable operation of the power grid.

At present, there are no relevant technical regulations for over-frequency generator tripping in China. Moreover, the over-frequency generator tripping program does not have its own fixed algorithm, which can only be set by multiple simulation analysis or engineering experience.

In fact, most of the literatures only focus on the optimization of the cut amount in the scheme, but not the frequency operation value. Wu Chen [8] considers that the hydropower unit has a wide adaptability to the frequency, and the speed protection value can be set to 75 Hz. However, the maximum allowable range of the turbine unit is only 51.5 Hz. It can be seen that the allowable frequency range of the hydropower unit is much wider than that the turbine unit. Wu Guoyang [9] believed that in practical operations, the manufactures and power plants often set the protection values of generator conservatively from the responsibility of protecting equipment. Yang Fan [10] considered that due to the weak link between the regional grid and main grid, the frequency fluctuation is large when the grid is disconnected. The high-frequency machine operation should not be premature. Zhou Lei[11] agrees that the hydropower units account for a large part of the Yunnan power grid, and the frequency range is relatively wide. The maximum operate frequency of the high frequency cutting in regional power network is even up to 55Hz. Therefore, the scheme of high frequency cutting machine with hydropower unit according to the frequency abnormal operating range of thermal power unit is unreasonable. Zhou Chuanmei, Sun Huadong, et al.'s [12~16] reach consensus that due to the flexibility of start and stop of hydropower units, the emergency measures are given priority of removing the hydropower generating units. When the capacity of hydropower units is insufficient, some thermal power units are removed. Chen Shuyong [16] proposals that the frequency limit of the hydropower plant can be increased appropriately or the cutting delay can be prolonged. This method can be used to avoid the over-cutting phenomenon. However, the paper does not explain why the frequency limits of hydropower units can be increased, nor suggest how much the frequency limit of hydropower generating units can be increased.

Focus on above problems, this paper first analyses the range of the abnormal operate values frequency and the reason why the turbine frequency limit can be increased. The research is based on the model of 10 machines and 39 nodes, and the first round of frequency operation value of the high-frequency cutting device of the hydropower unit is determined by simulation. By comparing with the conventional over-frequency generator tripping device, the over-frequency machine frequency operation value of the hydropower unit is determined. Finally, according to the conclusions above, the over frequency generator tripping scheme in the hydro-thermal power mixed power network is simulated and analysed, and the results are obtained.

2. Discussion on setting principle of over-frequency generator tripping for hydropower unit

At present, the national standard [17] clearly rules the scope of the abnormal frequency operation of turbine generators. As shown in table 1, the operating values of the over-frequency generator tripping and the UFLS device are usually based on this range.

Table.1 Abnormal frequency running time of steam turbine generator

Frequency range (Hz)	Cumulative allowable run time (min)	Allow run time each time (s)
$51.0<f\le51.5$	>30	>30
$50.5<f\le51.0$	>180	>180
$48.5<f\le50.5$	Continuous duty	Continuous duty
$48.0<f\le48.5$	>300	>300
$47.5<f\le48.0$	>60	>60
$47.0<f\le47.5$	>10	>20
$46.5<f\le47.0$	>2	>5

However, previous studies have shown [6] [18~19] that abnormal operating capacity of the hydro-generator is superior to the turbo-generator. The highest frequency can reach 75Hz, and the lowest value can reach 43.75Hz. Therefore, it's relatively conservative for the frequency value of the hydropower generator unit to be set according to the abnormal frequency allowable range of operation of the turbine unit.

3. Over-frequency protection of hydropower units

In china, there are clear national standards for the installation of over frequency protection devices: High frequency protection should be installed for 100MW or more turbo-generator or hydro-generator with load above rated frequency, protection operation and demagnetization or program tripping [20~21]. At the same time, Chinese standard [22] also stipulates that the high frequency protection device should cooperate with the over-frequency generator tripping device, and the protection value of the over-frequency protection should be higher than the last round of the over-frequency generator tripping device. In these regulations, the hydropower unit only needs to install the high frequency protection.
Nowadays, the set value of over frequency protection of hydropower plants in our country is usually determined based on operational experience. Once an accident occurs in the system, the load changes sharply and the unit frequency exceeds the range of 47~53Hz [23]. Therefore, in order to avoid the influence on the scheme of over frequency generator shedding for the main network, the power frequency protection is usually setting at 52~56Hz, and the time delay is 2~5s [24], some hydropower plants also have optimized the over-frequency protection and the optimization results are shown in the following table [23] [25~27].

Table.2 Optimal frequency of over-frequency protection for hydropower plants

Hydropower plant	Over frequency protection original frequency（Hz）	Over frequency protection optimization frequency（Hz）
Majitang Hydroelectric power plant	53.00	55.00
Nanjindu Hydropower Station	52.70	54.00
	52.48	54.00
Gaoliangjian Hydropower Station	50.50	55.00

All in all, the low-frequency tolerance of hydraulic turbine unit is strong. In the power system and the low-frequency operation range where the load is allowed, the operation of the hydropower unit is not limited, so the minimum threshold of the hydropower unit may not be considered. Because of the maximum frequency value of over-frequency protection for the hydropower unit is 55Hz, the frequency value of over-frequency generator tripping can be improved properly.

4. Optimization frequency operation value of over-frequency generator tripping for hydropower units

4.1. System overview and selection of the first round of frequency operation value

This paper uses the IEEE-recommended 10-machine 39-node model to study and analyze the action values of over-frequency generator tripping scheme, as shown in Fig 1. In the model, 10 hydropower units are installed and equipped with corresponding governors and excitation system devices.
Assuming that two lines of BUS16-BUS15 and BUS16-BUS17, if one of these is overhauled and the other has a three phase permanent short-circuit fault, or a three phase permanent short-circuit fault happens on both of the lines, it will result in the area near the BUS16 being disconnected from the main network and a power outgoing isolated network being formed. There are four hydraulic turbine

units in this power outgoing isolated grid, with a total active power output of 2350MW. Under different operating modes, the system will have an unbalanced power from 70MW to 920MW. If without a stable control measures, the system frequency deviation is shown in Fig. 2.

It can be seen from Fig.2 that in various operating modes of the power system, the system frequency can reach 61.5Hz when no frequency emergency control measures are taken. The highest frequency obtained with the system not destabilized is set as the threshold operation value for the first round of over-frequency generator tripping. The first round operation value of over-frequency generator tripping measures for hydropower units is 52.0 HZ.

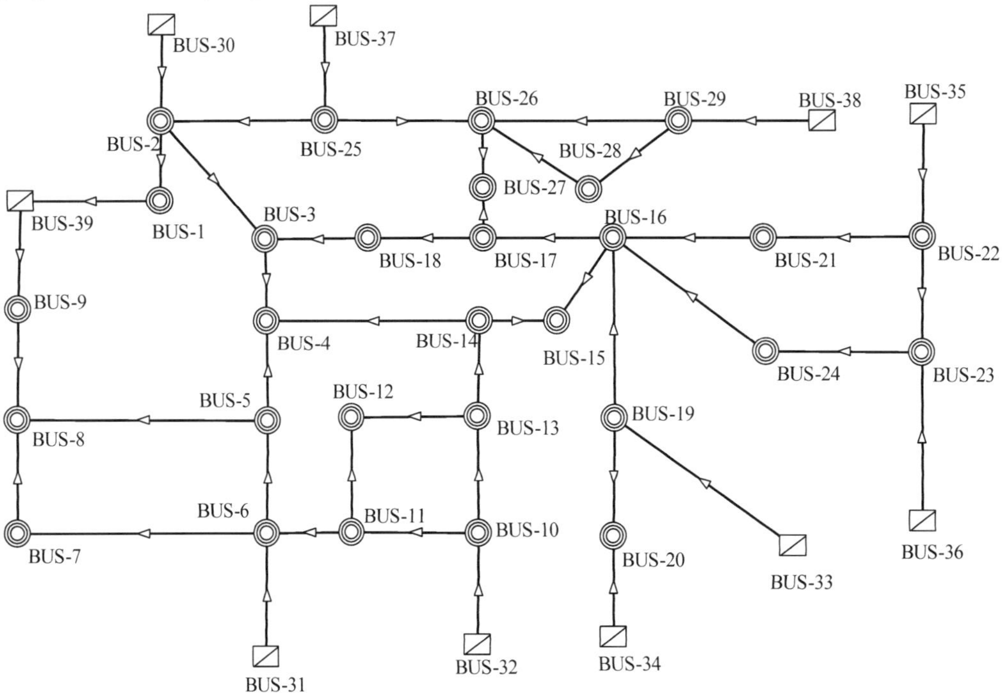

Fig. 1. 10-machine 39-node power flow diagram

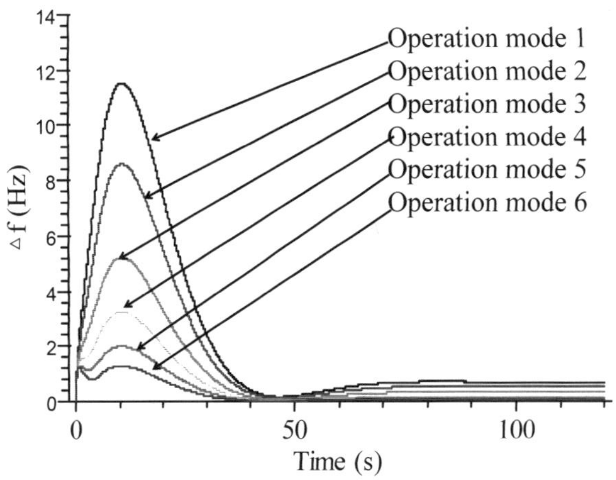

Fig. 2. System frequency deviation under different operation modes

4.2. Comparison and simulation of frequency operation values for Hydro-generator Units' over-frequency generator tripping

In addition to the over-frequency generator tripping, the system frequency emergency control measures should be provided with UFLS. According to the national standard and the system scale, there are five rounds of UFLS in this paper, with 0.2Hz difference and 0.2s delay per round. The first round of UFLS is 49Hz. Three rounds of over frequency generator tripping are set in the same way.

In general, the frequency operation value of the first round of over-frequency generator tripping in China is 50.8Hz, and the last round of frequency operation does not exceed 51.4Hz. From the 3.1 conclusion, it can be obtained that the first round value of over-frequency generator tripping operation is set as 52.0Hz. In this paper, the correctness of this conclusion is verified by the three modes of the operation mode 2, the operation mode 3 and the operation mode 6, with the first round of operation frequency being set to 50.8Hz, 52.0Hz and 52.5Hz respectively. The frequency deviation curves under different operating modes are shown in the following figures:

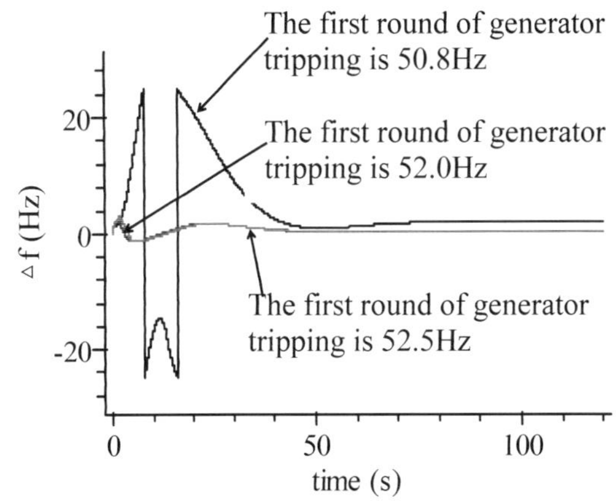

Fig 3 frequency deviation of operation mode 2

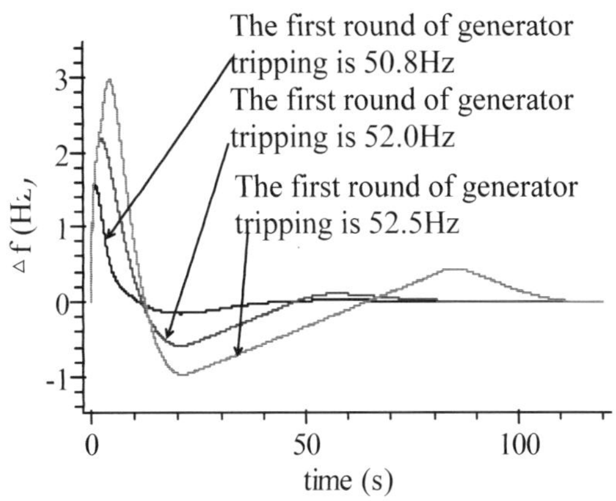

Fig 4 frequency deviation of operation mode 3

In Fig. 3, the frequency can't be controlled with all the three round of the over-frequency generator tripping device operating at 50.8Hz under operation mode 2. As the frequency fluctuation is too high, Triggering the action of the over-frequency protection device of the last unit, causing the frequency drops rapidly and removing all the loads, the system eventually loses the stability.. When the frequency value of the first round is 52.0Hz, the figure shows that the frequency quickly returns to stability after two round of over-frequency generator tripping and UFLS. When the first round of frequency optional value is 52.5Hz, the frequency fluctuation curve is similar to the one of 52.0Hz, while the fluctuation range is larger. Based on the above analysis, the first frequency operation value of the over-frequency generator tripping device should be selected as 52.0 Hz under this mode.

From Fig. 4, all three schemes go through only one round of over-frequency generator tripping and remove equal-capacity units. Comparing with the results in which the unit is removed at 50.8Hz and 52.0Hz, the values of stable frequency are almost same and the under frequency load shedding device is not triggered in both of the processes. Although the amplitude of the first frequency action is more obvious at the frequency of 52.0Hz than that at 50.8Hz during the frequency recovery process, it's still within the allowable range. If the unit is removed at 52.5Hz, it will trigger the first round of UFLS and take a longer time to make the frequency stable.

From Fig.5, the simulation system is operated in operation mode 6, with the first round of over frequency generator tripping device at 50.8Hz. According to the frequency curve, once the over frequency generator tripping device acts, the frequency drops sharply and causes in the action of UFLS. Because of the cyclic process, the system finally loses its stability. Besides, even if this system without any protective measures, the grid fault will not make the maximum frequency exceed 52.0 Hz. So, in the schemes with the over-frequency generator tripping devices set at 52.0Hz or 52.5Hz, the devices will not be triggered for the reason above. Only with the hydraulic turbine governing system and excitation system, the stability of the frequency can be maintained.

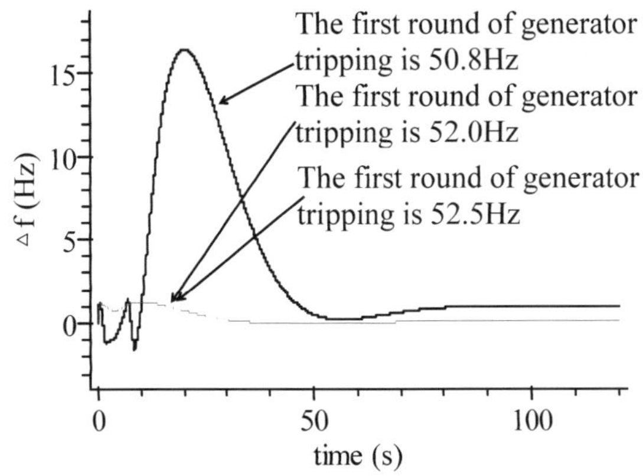

Fig 5 frequency deviation of operation mode 6

According to the simulation results, when the first round of the over-frequency generator tripping for the hydropower unit is set as 52.0Hz, the unnecessary load shedding or generator tripping can be avoided. It also ensures the safety and stability of the power grid and solves the problem that the mismatch between the over-frequency generator tripping and the UFLS which may lead to a blackout.

4.3. Over-frequency generator tripping scheme for priority removal of the thermal power units.

Considering that the tolerance of hydropower unit of frequency is stronger than thermal power units, the tripping frequency of hydropower units could be higher. The over-frequency generator tripping device should adopt the method of "removing the thermal power unit in priority" and reserve the function of frequency regulation by hydropower unit, which is more beneficial to the stability of the power system.

Based on the statistical data of the power system almanac in 2016, the thermal and hydropower units in the 10-machine 39-node model are set according to the current capacity of water and thermal power generators installed in China (about 76% :24%). In order to verify the correctness and feasibility of this method, it is still assumed the same faults above. But the BUS-36 unit in the power outgoing orphaned grid system was set as a hydropower unit, the other three units were set as thermal power units.

According to the traditional over-frequency generator tripping strategy, when the frequency reaches 50.8Hz, the first round of over-frequency generator tripping acts with a priority of removing the hydropower unit firstly. As the BUS-36 being removed, the frequency of the power system collapses. The frequency curve is shown in Fig.6.

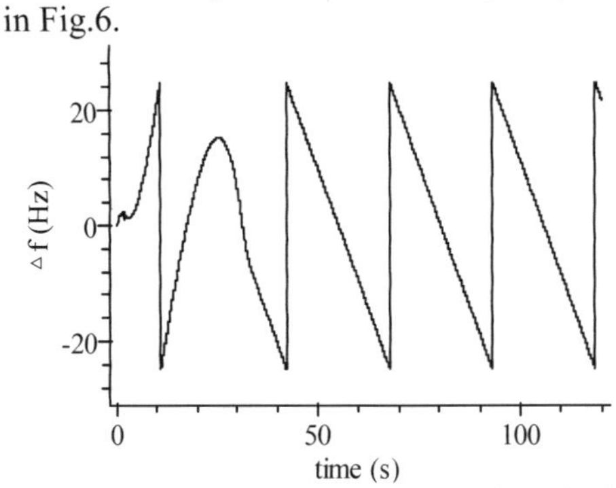

Fig 6 removing the hydropower units priority

If the thermal power unit in the system is set according to the traditional over-frequency generator tripping strategy, the first round of frequency operate value is 50.8Hz, while the operate value of hydropower is 52.0Hz. When the system frequency reaches 50.8Hz, the first round of over frequency generator tripping acts with a priority of removing the thermal power units firstly. The frequency curves are shown as follow.

When the system is operated in mode 3, its delivery power is 562.9MW, so the BUS-33 or BUS-34 thermal power unit can be considered when choosing to remove the thermal power unit. If the BUS-33 unit is cut off at the first round of the over frequency generator tripping at 50.8Hz, a round of UFLS will be triggered and the frequency will eventually return to be stable. If the BUS-34 unit is cut off at the first round of the over frequency generator tripping at 50.8Hz, it will not trigger the operation of the UFLS device and the frequency will also return to be stable.

From the comparison of the above figures, it can be concluded that the frequency regulation ability and tolerance ability of hydropower unit is superior to the thermal power unit. It's more beneficial to maintain the stability of frequency to remove the thermal power unit firstly.

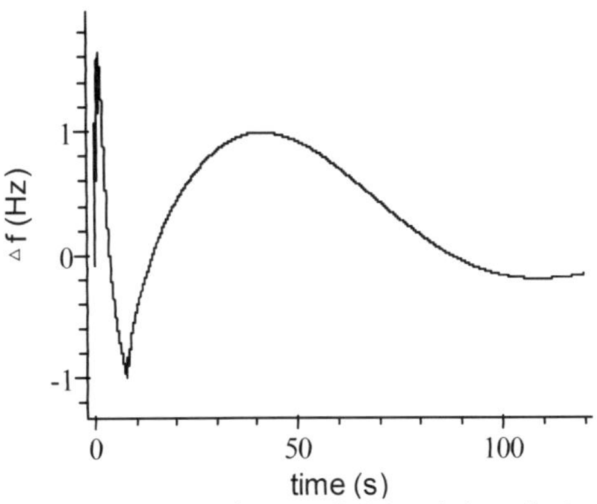

Fig 7 removing BUS-34 unit in priority

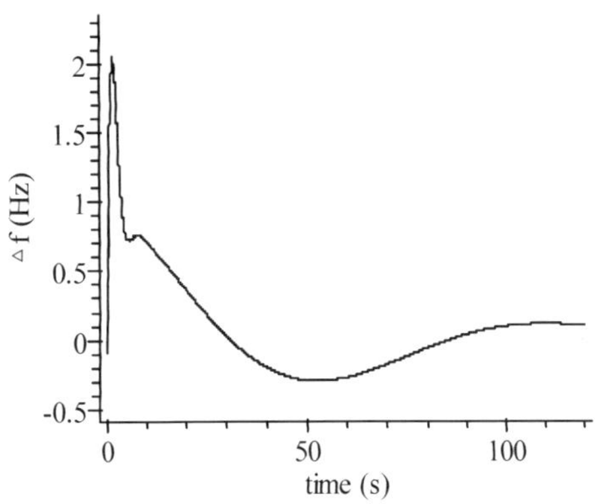

Fig 8 removing BUS-33 unit in priority

5. Conclusion

(1) The abnormal frequency is allowed to be set in a larger range, when the hydropower generating unit operates. If the frequency operate value of the over-frequency generator tripping is properly raised, the phenomenon of overcutting can be avoided as well as the repeated motion of over-frequency generator tripping and under frequency load reducer. It can prevent the power grid away from the collapse caused by power outages and make sure safety and stability of the grid operation.

(2) It ' s important to choose a suitable first round operation frequency of the over-frequency generator tripping. This frequency is set according to the maximum frequency value that ensures the system stability without any emergency frequency control measures.

(3) In the hydro-thermal power grid, when the frequency of the system is increased because of the excess power, the scheme of cutting a part of thermal power unit firstly is beneficial to stabilize the system frequency.

6. References

[1] Xinggang Wang, Peng Sun, Yingchun Qian. Research on High Frequency Problems of Xishuangbanna Power Grid. Yunnan Electric Power Technology.2010.J.38(2):5-8.

[2] Hao Jiang. Research on the security stability assessment and the measures of the isolated grid operation in Guizhou regional power grid. Guizhou University. MS thesis. China.2008.

[3] Yang Yu. Research on Power System Cascading outages and its relative issues. PhD dissertation. Department of Electrical Engineering, Zhejiang University, 2008.

[4] Ning Wang. Analysis on Problems in Relay Protection and Safety Automation Devices Viewed from “8.14” Blackout in USA & Canada .North China Electric Power. 2004. J. (7):50-53.

[5] Jianming Zhang, Tianjun Bi, Hui Liu. Review of frequency stability for isolated power system.

Power System Protection and Control. 2011. J. 39(11):149-154.
[6] Peigao Zhang, Xingyuan Li, Zheng Li. Research on frequency stability and control strategy in isolated power grid. Power System Protection and Control. 2012. J. 40(15):143-149
[7] Xinjian Cao, Yuhu Dai, Qiangsheng Lu. Study and Application of Qipanjing Stability Control System in isolated power system. Modern Manufacturing. 2016. J.(27):21-22.
[8] Tong Wu, Yigong Xie, Lingfang Li. Research and implementation of high frequency security and stability defense line in power grid. In: 2007 Yunnan Power Technology Forum Proceedings 2007.
[9] Guojun Wu, Xinli Song, Ping Yu. The Coordinated Control Principles for Large-scale Nuclear Power Units Protection and Automatic Safety Devices in the power grid. Electric Power System Automation. 2014. J. (3):178-183.
[10] Fan Yang, Zhiqiang Zhang, Shuqiang Zhao. The over-frequency generator tripping of regional power grid and cooperation with overspeed protection. Electric Power. 2013. J. 46(10):35-40.
[11] Lei Zhou, Dan Zhang, Fusuo Liu. Frequency Characteristics of Yunnan Power Grid and Corresponding Over-Frequency Generator Tripping Scheme after Asynchronous Interconnection. Southern Power System Technology . 2016. J. 10(7):17-23.
[12] Chuanmei Zhou, Bin Sun. Study on Security and Stability measures of the isolated grid operation of the Guizhou Province and area power grids. Power System Protection and Control. 2008. J. 36(19):29-32.
[13] Huadong Sun, Xuedong Wang, Shiying Ma. Measures to Improve System Security and Stability for Isolated Operation of Guizhou Main Power Grid and its Regional Power Networks. Power System Technology. 2008. J. 32(17):35-39.
[14] Zhuan Zhou, Xiuping Yao, Weiqing Wang. Stability Control Generator Tripping Based on Thermal/Hydro/Wind Bundled Power Transmission System. East China Electric Power. 2014. J. 42(2): 269-274.
[15] Wang Yizhen, Ma Shiying, Wang Qing, et al. Present Status and Development Trend of Research on Over-frequency in Isolated Power Grid. Power System Technology, 2012, 36(12):165-169.
[16] Shuyong Chen, Lin Zhu, Jian Ding. Impact of Grid-connected Wind Farms on High-frequency Generator Tripping in Isolated Power Grid. Power System Technology. 2012. J. 36(1):58-64.
[17] The grid operation code. GB/T 31464-2015 (2015).
[18] Lin Cai. Power System Frequency. China Electric Power Press, 1998
[19] Schleif F R. Study of interconnected power systems operation at below normal frequency. Research report. Schleif (Ferber R.), Denver, CO (USA) 1979.
[20] Technical code for relaying protection and security automatic equipment. China national standardization administration. GB/T 14285-2006 (2006)
[21] Guide on security and stability for power system. The national economic and trade commission of the People's Republic of China. DL.755-2001 (2001)
[22] Technical specification for grid-related protection of large generating unit. China national standardization administration. DL/T. 1309-2013 (2013)
[23] Zhiwen Hu. Technical Improvement of Shutdown which Caused by Over Frequency Protection of Units in Nanjindu Hydropower Station. Hunan Water Resources & Hydropower. 2005 J. (2):74-75.
[24] Zhengang Yang, Guocing Wu, Xiangdong Luo. The Safety-Stability Control Strategy and Tripping Principles for Northern Guangdong Power Grid. Southern Power System Technology. 2009. J., 3(2):15-17.
[25] Ping Peng. A study on the stable operation on tubular units in the isolated power grids. Hunan Electric Power. 1996. J. (3):43-47.
[26] Junhui Yang. Analysis of Dynamic Stability Operation of Nanjindu Hydropower Station. Hunan Water Resources and Hydropower. 1999.J.(5):43-45.
[27] Haibo Sun, Xingyun Wu. Application of High Frequency Relay in Control and Protection Circuit of Hydropower Station Generator Set. Jiangsu Water Resources. 2009.J.(10):23-24.

A New Principle of HVDC Protection Based on Two-terminal Electrical Quantities

Xinyue Zhang , Xiaoning Kang and Yagang Zhang

Xi'an Jiao Tong University, Xi'an, Shaanxi Province, People's Republic of China

E-mail: shakirayue@stu.xjtu.edu.cn

Abstract: This paper proposes a new principle of HVDC protection on two-terminal electrical quantities. Based on the analysis of short-circuit faults on different locations, it is found that as for current and voltage variation, there is a large difference between internal and external faults. According to this, the product of rectifier and inverter DC current variation can be used to distinguish the internal and external faults except for opposite pole line faults. In order to distinguish which DC line occurs short-circuit faults, bringing in rectifier and inverter DC voltage variation can achieve the target. This method identifies fault types quickly and has great tolerance of transition resistance. By building a HVDC model on PSCAD/EMTDC, the applicability of faults on different locations can be tested. The simulation result proves that the method has high reliability and can be applied to the whole line.

1. Introduction

Compared with AC transmission, DC transmission has an advantage over high capacity, long distance and asynchronous networking. Nowadays, Chinese HVDC transmission technology develops rapidly: China has already constructed 29 HVDC projects and put them into operation, including 7 UHVDC projects with ± 800 kV. These projects transmit energy from west regions to eastern load centers, which can effectively solve the problems of exploitation, transmission and consumption of hydropower, wind power, photovoltaic energy, producing remarkable economic and social profits.

At present, ABB's and SIEMENS's traveling wave protection schemes are applied widely for HVDC lines. Based on the catadioptric principle of traveling wave, these two schemes take traveling wave protection as main protection and take differential undervoltage protection and longitudinal differential protection as backup protections. The main criteria of traveling wave protection are DC voltage variation ΔU, DC voltage variation rate dU/dt and DC current variation ΔI. However, transition resistance has great influences on these criteria [1–4]. On condition that there is a large transition resistance in the DC system, relay protection device may reject act which will decrease reliability of the system. When DC line short-circuit fault occurs, DC current and voltage with plenty of fault information can be measured at the outlet of the rectifier and inverter. Therefore, DC current and voltage variation can be analyzed on different fault conditions [5–8]. This paper uses the product's positive and negative characteristics of DC current variation ΔI_R and ΔI_I on both sides as well as DC voltage variation ΔU_R and ΔU_I to distinguish internal and external faults. The simulation results show that the method can distinguish internal and external faults quickly and initiate protection device with internal faults. Besides, the method has great tolerance of transition resistance and can be applied widely.

2. DC current and voltage variation analysis with different locations's fault

As is shown in figure 1, bipolar HVDC transmission system includes rectifier(R side), inverter(I side), smoothing reactor, DC filter and DC transmission lines. u_{Rp} and u_{Rn} respectively represent rectifier side positive and negative pole voltage; i_{Rp} and i_{Rn} respectively represent rectifier side positive and negative pole current; u_{Ip} and u_{In} respectively represent inverter side positive and negative pole voltage and i_{Ip} and i_{In} respectively represent inverter side positive and negative pole current. The paper regulates that the positive direction of the current is from R side to I side.

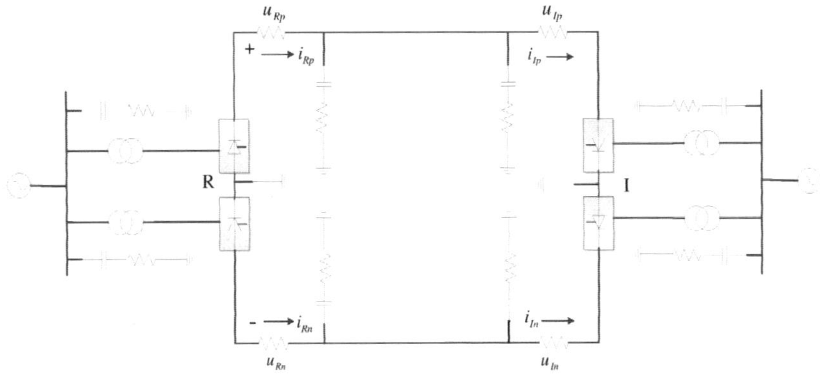

Figure 1. Bipolar HVDC system

Figure 2 shows that in the bipolar HVDC transmission system there are several locations that may occur short-circuit faults, including d_1, d_2, d_3 and d_4. Among these locations, d_1 represents internal fault and others represent external faults.

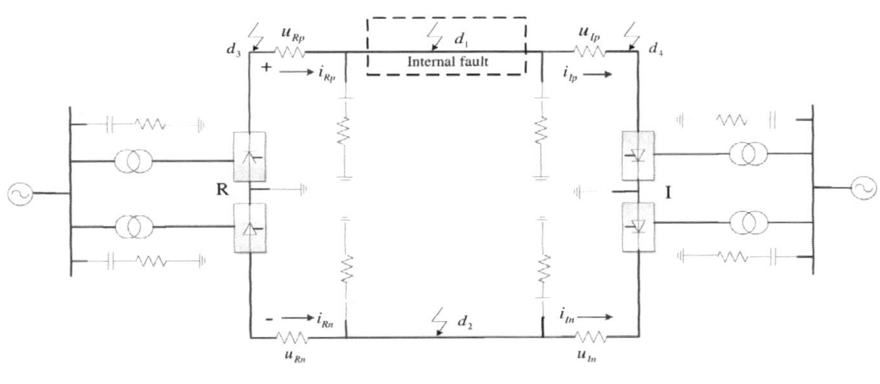

Figure 2. Faults on different locations in bipolar HVDC system

The paper aims to ensure that the protection device can correctly distinguish internal and external faults and protection wouldn't reject act with a large transition resistance. According to the analysis of DC current and voltage from rectifier and inverter side, the corresponding method can be proposed to distinguish different types of faults.

2.1. Electrical variations with internal faults

When DC line short-circuit fault occurs, it equals to add a reverse power U_f which has the same amplitude and opposite polarity compared with fault voltage according to the principle of superposition. As can be seen in figure 3, L_d represents smoothing reactor; Z_R and Z_I respectively represent the line equivalent impedance of the rectifier and inverter side, separated by the short circuit point and Z_F represents the equivalent impedance of the DC filter.

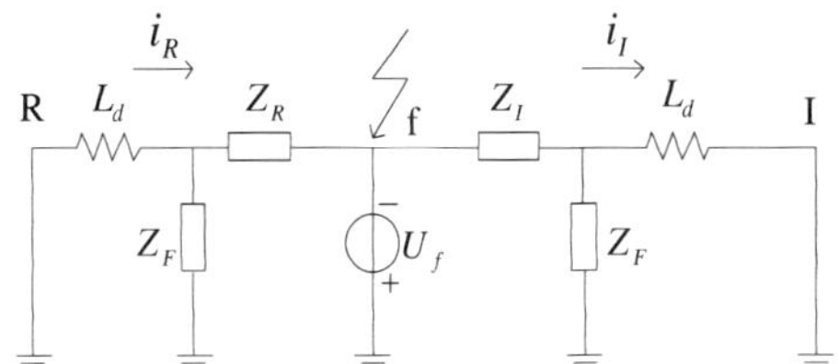

Figure 3. Equivalent circuit of internal fault

When f point occurs a short-circuit fault, reverse power U_f injects short-circuit current to rectifier and inverter side, respectively. The direction of short-circuit current rejecting to rectifier side is from rectifier to inverter side, so the rectifier DC current variation $\Delta I_R = I_R^{(1)} - I_R^{(0)} > 0$. However, the direction of short-circuit current rejecting to inverter side is from inverter to rectifier side, which means the inverter DC current variation $\Delta I_I = I_I^{(1)} - I_I^{(0)} < 0$. In addition, DC voltage variations from both sides decline sharply.

It can be inferred that:

$$\begin{cases} \Delta I_R > 0 \\ \Delta I_I < 0 \end{cases} \qquad \begin{cases} \Delta U_R < 0 \\ \Delta U_I < 0 \end{cases}$$

We can easily deduce that:

$$\Delta I_R \cdot \Delta I_I < 0$$

The interpole fault is the most serious fault belonging to internal faults which has the same changing direction of DC current and voltage as monopole faults. The only difference is that the variation of DC current and voltage is larger.

2.2. Electrical variations with external faults

2.2.1. Grounding short-circuit fault on the outlet of the rectifier. Figure 4 shows an equivalent circuit of grounding short-circuit fault on the outlet of the rectifier. Z_L represents the equivalent impedance of the DC transmission line.

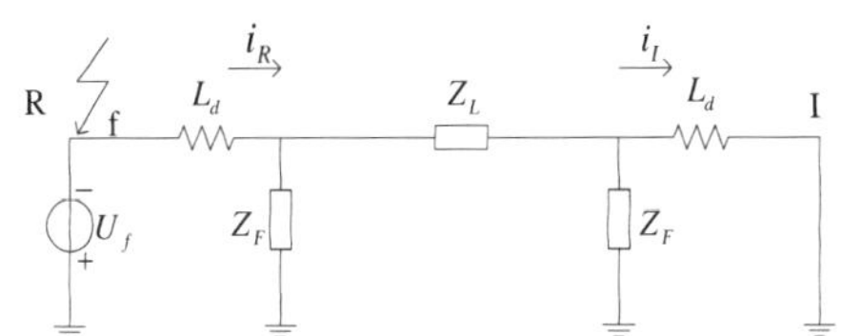

Figure 4. Equivalent circuit of grounding short-circuit fault on the outlet of the rectifier

When f point occurs short-circuit fault, reverse power U_f injects short-circuit current to rectifier and inverter side, respectively. The direction of short-circuit current rejecting to both sides is from inverter to rectifier side, so the rectifier DC current variation $\Delta I_R = I_R^{(1)} - I_R^{(0)} < 0$ and the inverter DC current variation $\Delta I_I = I_I^{(1)} - I_I^{(0)} < 0$.

It can be inferred that:

$$\begin{cases} \Delta I_R < 0 \\ \Delta I_I < 0 \end{cases}$$

We can easily deduce that:

$$\Delta I_R \cdot \Delta I_I > 0$$

2.2.2. Grounding short-circuit fault on the outlet of the inverter. Figure 5 shows an equivalent circuit of grounding short-circuit fault on the outlet of the inverter.

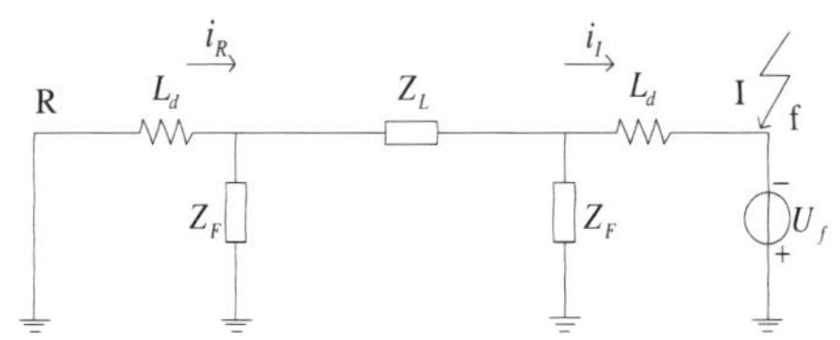

Figure 5. Equivalent circuit of grounding short-circuit fault on the outlet of the inverter

When f point occurs short-circuit fault, reverse power U_f injects short-circuit current to rectifier and inverter side, respectively. The direction of short-circuit current rejecting to both sides is from rectifier to inverter side, so the rectifier DC current variation $\Delta I_R = I_R^{(1)} - I_R^{(0)} > 0$ and the inverter DC current variation $\Delta I_I = I_I^{(1)} - I_I^{(0)} > 0$.

It can be inferred that:

$$\begin{cases} \Delta I_R > 0 \\ \Delta I_I > 0 \end{cases}$$

We can easily deduce that:

$$\Delta I_R \cdot \Delta I_I > 0$$

2.2.3. Grounding short-circuit fault on the opposite line.
Figure 6 illustrates that the only difference between grounding short-circuit fault on the opposite line and internal fault is the direction of the reverse power.

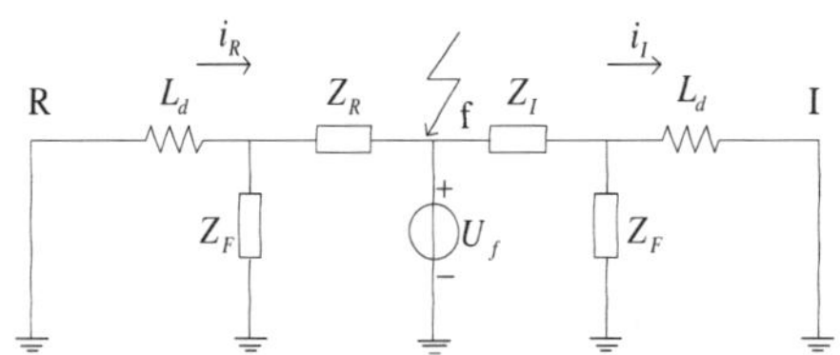

Figure 6. Equivalent circuit of grounding short-circuit fault on the opposite line

IOP Conf. Series: Journal of Physics: Conf. Series **1072** (2018) 012011 doi:10.1088/1742-6596/1072/1/012011

When f point occurs short-circuit fault, reverse power U_f injects short-circuit current to rectifier and inverter side on the opposite line, respectively. The direction of short-circuit current rejecting to the rectifier side is from inverter to rectifier side. When the current reaches DC bus on the rectifier side, part of the short-circuit current will flow from DC bus to the inverter side, so the rectifier DC current variation $\Delta I_R = I_R^{(1)} - I_R^{(0)} > 0$. On the contrary, the direction of short-circuit current rejecting to inverter side on the opposite line is from the rectifier to inverter side, and as soon as the short-circuit current reaches the DC bus on the inverter side, part of it will flow from DC bus to the rectifier side, which means the inverter DC current variation $\Delta I_I = I_I^{(1)} - I_I^{(0)} < 0$. Besides, DC voltage variations on both sides change slightly.

It can be inferred that:

$$\begin{cases} \Delta I_R > 0 \\ \Delta I_I < 0 \end{cases}$$

We can easily deduce that:

$$\Delta I_R \cdot \Delta I_I < 0$$

3. Protection principle

3.1. Protection setting process

Figure 7 displays the flow chart of HVDC protection based on two-terminal electrical quantities.

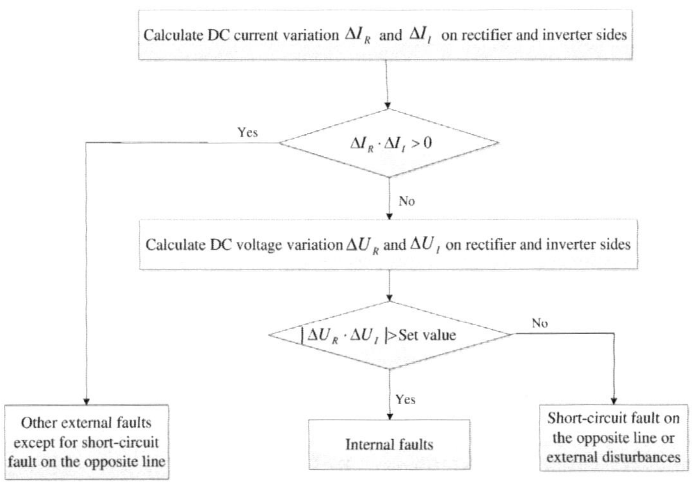

Figure 7. Flow chart of HVDC protection based on two-terminal electrical quantities

3.2. Protection setting calculation

When it comes to calculate the value of the DC voltage variation $|\Delta U_R \cdot \Delta U_I|$, avoiding the maximum value of the fault on the opposite DC line should be considered and calculated as equation (1). The corresponding sensitivity check can be calculated as equation (2).

$$|\Delta U_R \cdot \Delta U_I|_{set} = K_{rel} \cdot |\Delta U_R \cdot \Delta U_I|_{out} \tag{1}$$

$$K_{sen} = \frac{|\Delta U_R \cdot \Delta U_I|_{\min}}{|\Delta U_R \cdot \Delta U_I|_{set}} \tag{2}$$

where $|\Delta U_R \cdot \Delta U_I|_{set}$ represents the protection set value of DC voltage variation, $|\Delta U_R \cdot \Delta U_I|_{\min}$ represents the minimum value of internal faults, and K_{rel} represents confidence coefficient.

4. Simulation analysis

4.1. Simulation of protection principle

By means of PSCAD/EMTDC, we can get diagrams of DC current and voltage on both sides. Suppose that the grounding short-circuit fault occurs on the middle of the DC transmission line: the fault occurs at 0.5s, fault duration is 0.1s and the simulation time is 2s. Figure 8 shows DC current and voltage with internal faults; Figure 9 shows DC current with grounding short-circuit fault on the outlet of the rectifier; Figure 10 shows DC current with grounding short-circuit fault on the outlet of the inverter; Figure 11 shows DC current and voltage with grounding short-circuit fault on the opposite DC line.

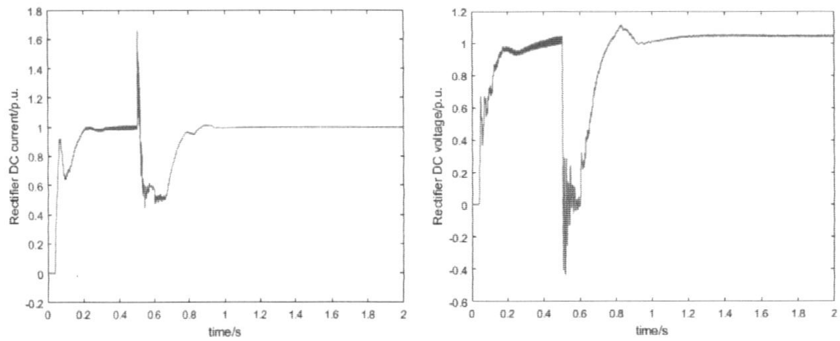

Figure 8(a). Rectifier DC current and voltage

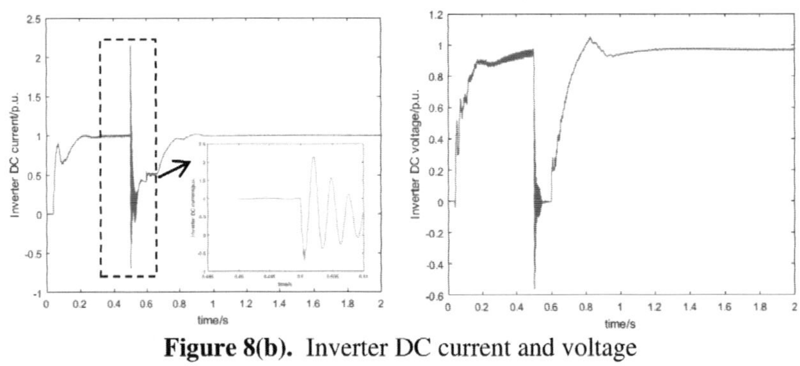

Figure 8(b). Inverter DC current and voltage

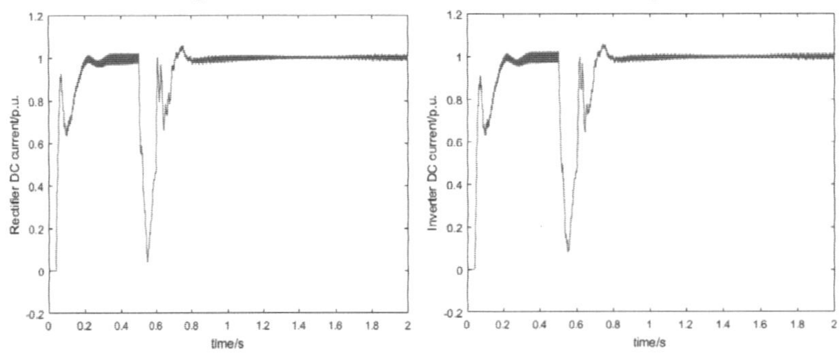

Figure 9. Rectifier and inverter DC current

IOP Conf. Series: Journal of Physics: Conf. Series **1072** (2018) 012011 doi:10.1088/1742-6596/1072/1/012011

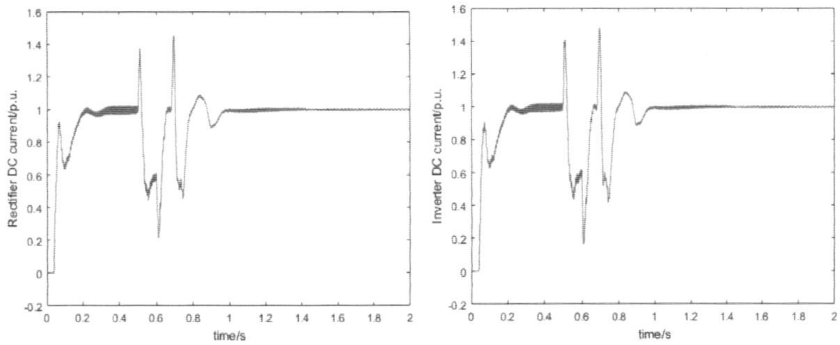

Figure 10. Rectifier and inverter DC current

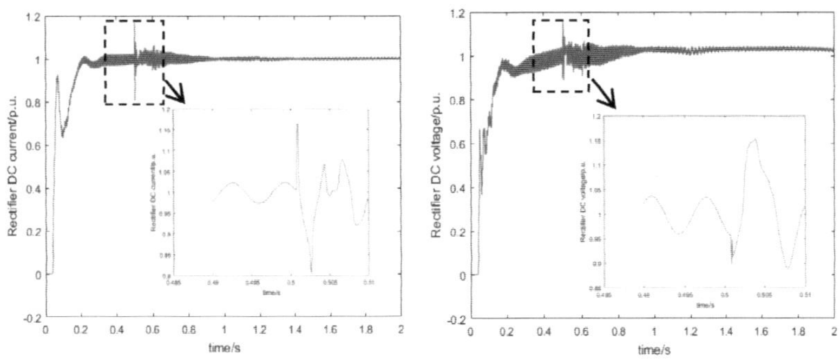

Figure 11(a). Rectifier DC current and voltage

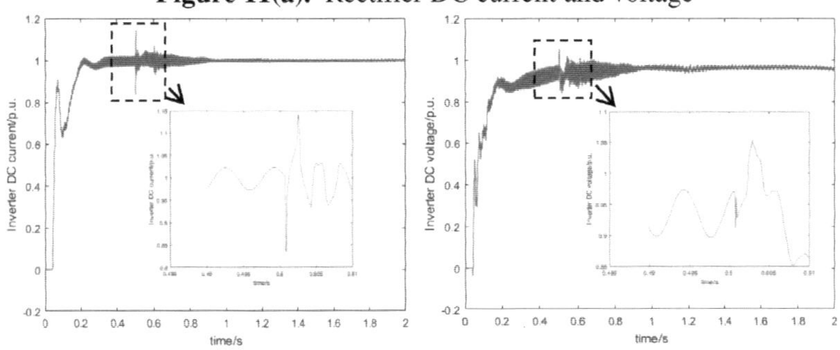

Figure 11(b). Inverter DC current and voltage

The above-mentioned information will be gathered in table 1.

Table 1. Variation of two-terminal electrical quantities in different fault situations

Fault type	ΔI_R	ΔI_I	$\Delta I_R \cdot \Delta I_I$	$\lvert \Delta U_R \cdot \Delta U_I \rvert$
Internal fault	>0	<0	<0	Larger
Grounding short-circuit fault on the outlet of the rectifier	<0	<0	>0	\
Grounding short-circuit fault on the outlet of the inverter	>0	>0	>0	\

Grounding short-circuit fault on the opposite line	>0	<0	<0	Smaller

Based on the analysis, we can get the information that when external faults except for grounding short-circuit fault on the opposite line occur, the relay protection device can distinguish internal and external faults through the positive and negative characteristics of $\Delta I_R \cdot \Delta I_I$. And the transition resistance won't influence the process of judgement. However, when grounding short-circuit fault on the opposite line occurs or there exists external disturbance in the system, judging the positive and negative characteristics of $\Delta I_R \cdot \Delta I_I$ cannot distinguish internal and external faults. The simulation results illustrate that compared with internal faults, DC voltage variations on both sides are much smaller. Therefore, the value of $|\Delta U_R \cdot \Delta U_I|$ can be used to choose the fault line. As for it, the greatest advantage is that protection device can correctly operate even though there is a large transition resistance on the fault point or external disturbance affects the system.

4.2. Protection sensitivity check

By means of PSCAD/EMTDC, we can build bipolar HVDC transmission system of ± 500 kV with rated transmission power of 1000MW and 500-kilometer-long line. Fault points will be set in the middle of positive and negative line, respectively. In the process of simulation, sampling frequency is 10kHz. In addition, the example requires that $K_{rel} = 1.2$ and the set value of $|\Delta U_R \cdot \Delta U_I|$ should meet the demand that $K_{sen\text{-}\min} = 2.5$. According to the simulation results, when the grounding short-circuit occurs on the inverter side of the opposite line, the measured value of $|\Delta U_R \cdot \Delta U_I|$ reaches its maximum. Since it takes some time for traveling wave to reach the middle of the line, protection device will operate in 1.5 ms. The values of $|\Delta U_R \cdot \Delta U_I|$ with internal and external faults are shown as table 2.

Table 2 The value of $|\Delta U_R \cdot \Delta U_I|$ of internal and external faults

Fault location	Internal fault(d_1)	External fault(d_2)
$|\Delta U_R \cdot \Delta U_I|$(p.u.)	1.7191	0.0987

With equation (1) and (2), we can get:

$$|\Delta U_R \cdot \Delta U_I|_{set} = K_{rel} \cdot |\Delta U_R \cdot \Delta U_I|_{out} = 1.2 \times 0.0987 = 0.1184$$

$$K_{sen} = \frac{|\Delta U_R \cdot \Delta U_I|_{\min}}{|\Delta U_R \cdot \Delta U_I|_{set}} = \frac{1.7191}{0.1184} = 14.5$$

4.3. The effects of transition resistance

When external faults except for grounding short-circuit fault on the opposite line occur, the protection device can distinguish internal and external faults through the positive and negative characteristics of $\Delta I_R \cdot \Delta I_I$. And the transition resistance won't influence the process of judgement. Even if there exists external disturbance in the system, measuring whether the value of $|\Delta U_R \cdot \Delta U_I|$ is larger than set value can filter the disturbance and it also can judge which line occurs short-circuit faults. In this process, transition resistance will slightly influence the protection. On this condition, the measured DC voltage variation with grounding short-circuit fault on the opposite line is much smaller than internal faults. Thus, the protection can act correctly. The protection actions of internal and external faults under different transition resistances are shown in table 3.

Table 3. Protection actions of internal and external faults under different transition resistances

Fault point	Transition resistance(Ω)	$\lvert \Delta U_R \cdot \Delta U_I \rvert$	Judging results
d_1	0.1	2.39	Internal fault
	100	1.0878	Internal fault
	800	0.1628	Internal fault
d_2	0.1	0.0255	External fault
	100	0.0165	External fault
	800	0.0066	External fault

4.4. The effects of control system

In general, the control system works in 5~6ms after a fault occurs. By analyzing the simulation results, we know that the protection device may have mistaken actions in a period of time when the control system begins to work until current and voltage reach to their rated value. In this case, the protection scheme can be the main protection for 5~6ms after the protection device begins to operate. Then the protection will be out of service 5~6ms later and backup protections containing differential undervoltage protection and longitudinal differential protection will work as main protections until DC current and voltage reach to their rated value.

4.5. The effects of fault locations

It is likely that the protection actions may be affected by fault locations. Therefore, it is necessary to analyze the effects of fault locations. Table 4 displays protection actions of internal and external faults under different locations.

Table 4 Protection actions of internal and external faults under different locations

Fault point	Distance of fault point to rectifier side(km)	$\lvert \Delta U_R \cdot \Delta U_I \rvert$	Judging results
d_1	10	1.8340	Internal fault
	250	2.39	Internal fault
	490	1.7191	Internal fault
d_2	10	0.0536	External fault
	250	0.0255	External fault
	490	0.0987	External fault

5. Conclusion

This paper proposes a new principle of HVDC protection based on two-terminal electrical quantities. By judging the positive and negative characteristics of $\Delta I_R \cdot \Delta I_I$, we can distinguish internal and external faults except for grounding short-circuit fault on the opposite line. However, as grounding short-circuit fault on the opposite line occurs, we can use the value of $|\Delta U_R \cdot \Delta U_I|$ to choose the fault line. Above all, the scheme is still available with a large transition resistance. Compared with widely applied DC protections, the scheme only makes use of two-terminal currents and voltages and the protection acts quickly. Besides, it has strong anti-interference ability with high reliability.

6. References

[1] X. Liu, A.H. Osman and O.P. Malik 2009 Hybrid traveling wave/boundary protection for monopolar HVDC line *IEEE Trans. Power Deliv.* **24** 569-78
[2] M. Xu, Z.X. Cai and K.L. Han et al. 2013 A sensitive and high-speed traveling wave protection scheme for HVDC transmission line *Int. Trans. Electr. Energy Syst.* **23** 1304-16
[3] Y. Zhang, N.L. Tai and B. Xu 2012 Fault analysis and traveling-wave protection scheme for bipolar HVDC lines *IEEE Trans. Power Deliv.* **27** 1583-91
[4] H. Ha, Y. Yu and R. Yi 2010 Novel scheme of traveling wave based differential protection for bipolar HVDC transmission lines *Proc. Int. Conf. Power Systems Technology* 1-5
[5] K. Han, Z. Cai and X.U. Min et al. 2013 Dynamic characteristics of characteristic parameters of traveling wave protection for HVDC transmission line and their setting *Power Syst. Technol.* **1** 255-60
[6] G.B. Song, X. Chu and S.P. Gao et al.2015 A new whole-line quick-action protection principle for HVDC transmission lines using one-end current *IEEE Trans. Power Deliv.* **30** 599-607
[7] S.L. Chen, H.C. Shu and C.H. Wan et al. 2013 The principle of single-ended transient based voltage protection for UHVDC transmission line *Power Syst. Prot. Control* **41** 26-31
[8] F. Kong, Z.G. Hao and S. Zhang et al. 2014 Development of A novel protection device for bipolar HVDC transmission lines *IEEE Trans. Power Deliv.* **29** 2270-78
[9] S.X. Luo, D.X. Dong and S.X. Shi et al. 2016 A directional protection scheme for HVDC transmission lines based on reactive energy *IEEE Trans. Power Deliv.* **31** 559-67
[10] G.B. Song, X. Chu and S. Gao et al. 2013 A whole-line quick-action protection principle for HVDC transmission lines using one-end current of DC-filters *Proc. CSEE* **33** 120-26
[11] J. Suonan, S. Gao and G. Song et al. 2010 A novel fault-location method for HVDC transmission lines *IEEE Trans. Power Deliv.* **25** 1203-09

Acknowledgements

I would like to express my gratitude to all those who helped me during the writing of this thesis. My deepest gratitude goes first and foremost to Professor Kang Xiaoning, my supervisor, for his constant encouragement and guidance. Last but not the least, my gratitude also extends to my family who have been assisting, supporting and caring for me all of my life.

IOP Conf. Series: Journal of Physics: Conf. Series **1072** (2018) 012012 doi:10.1088/1742-6596/1072/1/012012

Complexity Evaluation Model for Smart Grid Management Information System Based on Entropy Theory

Yifu Mo, Yongjun Zhang

School of Electric Power, South China University of Technology, Wushan Road, Guangzhou, China

578299002@qq.com

Abstract. System complexity evaluation is an important part in the process of smart grid management information system construction, but there are seldom researches in this area. By analyzing the complex features of smart grid management information system, this paper establishes a complexity evaluation index system from three aspects of primary station, communication network and intelligence terminal device. Besides, a complexity evaluation method based on entropy theory is proposed to quantify the complexity factors. And complex networks space theory is introduced to establish the complexity evaluation model for the whole system,which is used to provide guidance and support for system design, optimization and maintenance. Finally, the efficience of this model is verified by a real case of smart grid management information system in a city.

1. Introduction

Smart grid management information system(SGMIS) is an important platform for two-way interaction between power generation and utilization. With the rapidly expanding of the smart grid construction, the complexity of SGMIS has dramatically increased. It' s necessary to quantify the complexity of smart grid management information system in order to provide guidance for system design and optimization as well as maintenance service pricing.

In the system complexity science, entropy has become an important indicator to quantify the system complexity since Shannon proposed this concept to describe the system information content, and it is widely used in many areas such as analysis of multichannel data [1], Alzheimer' s Disease evaluation [2], quantifying the stock market inefficiency [3], detection of driver fatigue [4], etc.

Focusing on the complexity of SGMIS, architectures for SGMIS are proposed to provide guidance for the complexity estimate [5-7]. The complex features of SGMIS are analysed based on complexity adaptive system' s characteristics [8]. However, this paper just rests on the description of the complex features of SGMIS instead of quantifying the system complexity.

In general, the complex features of SGMIS have still not been understand and studied entirely [9], and the complexity assessment on SGMIS remains to be investigated.

Considering the entropy theory has been applied maturely. Firstly, this paper analyses the complex features of SGMIS, and establishes a complexity evaluation index system from three aspects of primary station, communication network and intelligence terminal device. Secondly, a complexity evaluation method based on entropy theory is proposed to quantify the complexity factors. Thirdly, the complex networks space theory is introduced to establish the complexity evaluation model for the whole system,which is used to provide guidance and support for system design, optimization and maintenance. Finally, the efficience of this model is verified by a real case of smart grid management information system in a city.

2. Complexity evaluation index system of SGMIS

Integrating the whole process of usage of electricity, SGMIS information flow exists in power generation, transmission, substation, distribution, consumption and so on five links, and contains the data from SCADA(Supervisory Control And Data Acquisition), EMS(Energy Management Systems) and etc. The framework of SGMIS is composed of three parts: primary station, communication network and intelligence terminal device.

According to the framework of SGMIS, this paper evaluates the its complexity from three aspects of primary station, communication network and intelligence terminal device. The evaluation index system of SGMIS is shown in Fig. 1.

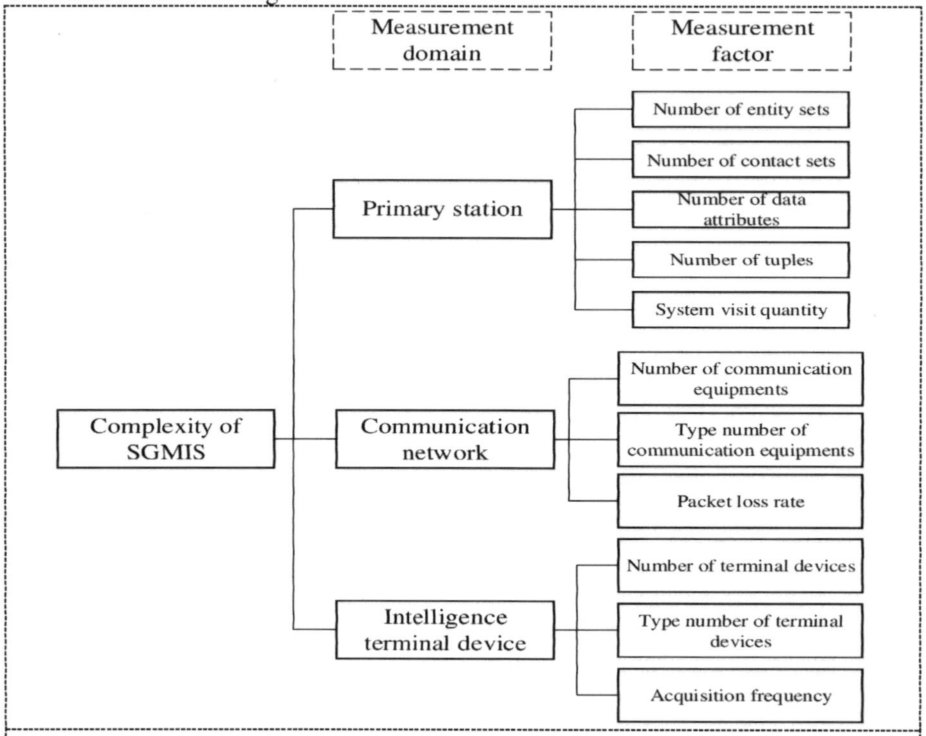

Figure 1. Evaluation index system of complexity about smart grid information management system

3. Complexity evaluation method based on entropy theory

Presented by Shannon, entropy is usually used to measure the amount of information. In the study of system complexity, entropy can be also used to represent the uncertainty of system structure and state. The bigger the entropy is, the greater the uncertainty of system structure becomes.

Assuming that the system has n states, and the structure entropy of this system is shown in (1), where the probability of State i is P_i.

$$H=-\sum_{i=1}^{n} P_i \log_2 P_i \tag{1}$$

3.1. Calculation methods of structure entropy in primary station layer

3.1.1. Number of entity sets

Generally speaking, entities with same attribute are often classed as a set. Squares symbolize entities while rectangles with a name symbolize entity sets in E-R diagram. The structure entropy of entity set is shown in (2), where the number of entity sets is n_e.

$$H_{ne}=n_e[-\frac{1}{n_e}\log_2\frac{1}{n_e}]=-\log_2\frac{1}{n_e} \quad (2)$$

3.1.2. Number of contact sets

Referring to the relation between entities, contacts with same attribute are often classed as a contact set, and it is symbolized by a diamond in E-R model. The structure entropy of contact set is shown in (3), where the number of contact sets is n_r.

$$H_{nr}=n_r[-\frac{1}{n_r}\log_2\frac{1}{n_r}]=-\log_2\frac{1}{n_r} \quad (3)$$

3.1.3. Number of attributes

Attribute is the nature of each object in the entity sets or contact sets. The number of attributes is equal to the number of ellipses in the E-R model. Assuming that the number of entity sets and contact sets is n_p, where $n_p=n_e+n_r$. The structure entropy of attribute is shown in (4), where the number of attributes of entity set i or contact set i is n_i.

$$H_{np}=-\sum_{i=1}^{n_p}\log_2\frac{1}{n_i} \quad (4)$$

3.1.4. Number of tuples

Corresponding to one row in data sheet, each tuple has a unique number. Assuming that there are n data sheets, the structure entropy of data sheet i is shown in (5), where the number of tuples in data sheet i is n_{ti}.

$$H_{nt}=\sum_{i=1}^{n}n_{ti}(-\frac{1}{n_{ti}}\log_2\frac{1}{n_{ti}})=-\sum_{i=1}^{n}\log_2\frac{1}{n_{ti}} \quad (5)$$

3.1.5. System visit quantity

After the modeling analysis of static complexity in primary station layer, dynamic complexity can be described by system visit quantity. Supposing that the number of user actions during a observing period is n_v, the structure entropy of system visit quantity is shown in (6).

$$H_{nv}=n_v(-\frac{1}{n_v}\log_2\frac{1}{n_v})=-\log_2\frac{1}{n_v} \quad (6)$$

3.2. Calculation methods of structure entropy in communication network layer

3.2.1. Number of communication equipments

The structure entropy of the number of communication equipments is shown in (7), where the number of communication equipments is n_c.

$$H_{nc}=n_c(-\frac{1}{n_c}\log_2\frac{1}{n_c})=-\log_2\frac{1}{n_c} \quad (7)$$

3.2.2. Type number of communication equipments

The structure entropy of the type number of communication equipments is shown in (8), where the type number of communication equipments is n_{tc}.

$$H_{tc}=n_{tc}(-\frac{1}{n_{tc}}\log_2\frac{1}{n_{tc}})=-\log_2\frac{1}{n_{tc}} \quad (8)$$

3.2.3. Packet loss rate

The structure entropy of packet loss rate is shown in (9), where the packet loss rate of type i communication equipments is r_i.

$$H_{\mathrm{r}}=\sum_{i=1}^{n_{\mathrm{ta}}} r_i(-\frac{1}{r_i}\log_2\frac{1}{r_i}) \tag{9}$$

3.3. Calculation methods of structure entropy in intelligence terminal device layer

3.3.1. Number of terminal devices

Supposing that the number of terminal devices is n_{a}, the structure entropy of it is given in (10).

$$H_{\mathrm{na}}=n_{\mathrm{a}}(-\frac{1}{n_{\mathrm{a}}}\log_2\frac{1}{n_{\mathrm{a}}})=-\log_2\frac{1}{n_{\mathrm{a}}} \tag{10}$$

3.3.2. Type number of terminal devices

Supposing that the type number of terminal devices is n_{ta}, the structure entropy of it is given in (11).

$$H_{\mathrm{ta}}=n_{\mathrm{ta}}(-\frac{1}{n_{\mathrm{ta}}}\log_2\frac{1}{n_{\mathrm{ta}}})=-\log_2\frac{1}{n_{\mathrm{ta}}} \tag{11}$$

3.3.3. Acquisition frequency

The structure entropy of acquisition frequency is shown in (12), where the acquisition frequency of type i terminal devices is f_i.

$$H_{\mathrm{f}}=\sum_{i=1}^{n_{\mathrm{ta}}} f_i(-\frac{1}{f_i}\log_2\frac{1}{f_i}) \tag{12}$$

4. Complexity evaluation model based on complex networks space theory

Introducing the complex networks space theory, a complexity evaluation model is built in order to analyse the system complexity ponderance and resultant.

4.1. complexity ponderance evaluation model

According to the complexity features of SGMIS, management energy [10] is introduced to describe the complexity and administrative cost of each layer.Assuming that $\boldsymbol{e}_i$ is a an n-vector, and $\boldsymbol{e}_i=(e_{i1}, e_{i2}, \ldots\ldots, e_{in})$, where $\boldsymbol{e}_{in}$ represents the structure entropy of layer i in the system. On the basis of management energy and Euclidian distance in the complex networks space, system complexity ponderance can be obtained in (13).

$$w_i = e_{in}\sqrt{\|\boldsymbol{e}_i\| - e_{in}^2} \tag{13}$$

4.2. complexity resultant evaluation model

An united structure entropy space is built in order to get the system complexity resultant. Supposing that complexity vector space $\boldsymbol{E}$ includes n vectors with m dimensions, where m represents the number of layers in the system and n represents the number of complexity evaluation indexes. Complexity vector $\boldsymbol{e}_i$ is defined as the element of $\boldsymbol{E}$ and $\boldsymbol{e}_i \in \boldsymbol{H}^n$. Then the structure entropy matrix can be obtained.

$$\boldsymbol{M}=\begin{bmatrix}\boldsymbol{e}_1\\ \vdots\\ \boldsymbol{e}_i\\ \vdots\\ \boldsymbol{e}_m\end{bmatrix}=\begin{bmatrix}e_{11} & e_{12} & \cdots & \cdots & e_{1n}\\ \vdots & \ddots & \cdots & \cdots & \vdots\\ e_{i1} & e_{i2} & \ddots & \cdots & e_{in}\\ \vdots & \vdots & \vdots & \ddots & \vdots\\ e_{m1} & e_{m2} & \cdots & \cdots & e_{mn}\end{bmatrix}$$

The distance between two vectors in matrix $\boldsymbol{M}$ is given in (14).

$$\|\boldsymbol{e}_{i+1}-\boldsymbol{e}_i\| = \|e_{(i+1)1}-e_{i1}, e_{(i+1)2}-e_{i2}, \cdots, e_{(i+1)n}-e_{in}\| \tag{14}$$

A complexity vector space with n dimensions is defined after combining m vectors: $\boldsymbol{e}_1\times\boldsymbol{e}_2\times\ldots\ldots\times\boldsymbol{e}_m$. Assuming that $\boldsymbol{\Phi}$ is a image and tensor in $\boldsymbol{H}$, and $\boldsymbol{\Phi}$: $\boldsymbol{e}_1\times\boldsymbol{e}_2\times\ldots\ldots\times\boldsymbol{e}_m\rightarrow\boldsymbol{H}$. $\boldsymbol{\Phi}$ is defined as $T_m(\boldsymbol{H})$ and a vector space in $\boldsymbol{H}$.

0 elements can be used to supplyment the vector with less columns while the vectors in $\boldsymbol{M}$ don't have the same number of columns, and the tensor is described: $\|\boldsymbol{\Phi}_i\|=\|\boldsymbol{e}_i\|$, $\|\boldsymbol{\Phi}_i\|$ refers to the united structure entropy of layer i.

When the vectors in $\boldsymbol{M}$ have the same attribute, the complexity resultant of the whole system is given in (15).

$$\|C\|_{\mathrm{E}}=\sum_{i=1}^{m-1}\|\triangleleft(\boldsymbol{e}_{i+1})-\triangleleft(\boldsymbol{e}_i)\|=\sum_{i=1}^{m-1}\|\boldsymbol{e}_{i+1}-\boldsymbol{e}_i\| \quad (15)$$

In (15), $\|\boldsymbol{e}_{i+1}-\boldsymbol{e}_i\|$ is the complexity information that layer i+1 generates after layer i, and m is the total number of system layers.

Conversely, the complexity resultant of the whole system is given in (16).

$$\|C\|_{\mathrm{T}}=\sum_{i=1}^{m}\|\triangleleft(\boldsymbol{e}_i)\|=\sum_{i=1}^{m}\|\boldsymbol{e}_i\| \quad (16)$$

While both the vectors with the same attributes and those with different attributes exit simultaneously, the complexity information of the vectors with the same attributes can be calculated by (15), whereas the complexity information of the vectors with different attributes can be calculated by (16). Finally, the complexity resultant of the whole system is the sum of the two above-mentioned results.

5. Case study

5.1. Introduction of SGMIS in a city

On the basis of the various electric power load control systems built before, a city power supply bureau establishes a SGMIS through energy management system(EMS), distribution management system(DMS) and so on. Statistic of complexity data is shown in Table 1.

Table 1. Statistic of complexity data

Mesurement domain	Mesurement factor	Statistic	Structure entropy calculation
Primary station	Number of entity sets	96	6.5850
	Number of contact sets	150	7.2288
	Number of data attributes	——	103.3971
	Number of tuples	——	328.7741
	System visit quantity	7293881	22.7983
Communication network	Number of communication equipments	266193	18.0221
	Type number of communication equipments	24	4.5850
	Packet loss rate	——	28.0494
Intelligence terminal device	Number of terminal devices	6615248	22.6574
	Type number of terminaldevices	12	3.5850
	Acquisition frequency	——	55.7879

5.2. Calculation analysis

The complexity ponderances can be calculated by (13), and the united structure entropy of each layer can be obtained according to $\|\triangleleft_i\|=\|\boldsymbol{e}_i\|$. The result is given in Table 2.

Table 2. Statistic of complexity data

	complexity ponderance	united structure entropy
Primary station	7635.843	335.706
Communication network	521.6113	33.6539
Intelligence terminal device	1279.7324	60.32

Because the complexity vectors have different attributes, the complexity resultant is calculated by (16).

$$\|C\|_{\mathrm{T}} = \sum_{i=1}^{3}\|\sigma(e_i)\| = \sum_{i=1}^{3}\|e_i\|$$
$$= 335.706 + 33.6539 + 60.32 = 429.68$$

From the complexity ponderances, $w_1 > w_3 > w_2$, the primary station is the most complex part. Meanwhile, the increasing complexity in communication network layer and intelligence terminal device layer makes it essential to define relative standards of the devices, in order to standardize the management and reduce the management cost.

6. Conclusion

System complexity evaluation is a precondition for estimating the maintenance cost. This paper proposes the complexity evaluation model based on complex networks space theory, broadening a new horizon and leads to the complexity evaluation.

According to the complexity ponderances w_1, w_2, w_3, and complexity resultant $\|C\|_{\mathrm{T}}$, the system complexity can be evaluated locally or generally. The bigger the complexity resultant is, the more complex the system is. It provides a guidance for design scheme optimization, which is very practical in application. And it will be more useful in wider area if corresponding software could be developed.

7. References

[1] Ahmed M U and Mandic D P 2011 Multivariate multiscale entropy: a tool for complexity analysis of multichannel data *Physical Review E Statistical Nonlinear & Soft Matter Physics* 84(6 Pt 1):061918.

[2] Morabito F C, Labate D, Foresta F L, et al 2012 Multivariate Multi-Scale Permutation Entropy for Complexity Analysis of Alzheimer's Disease EEG *Entropy* 14(7):1186-1202.

[3] Zunino L, Zanin M, Tabak B M, et al 2012 Complexity-entropy causality plane: A useful approach to quantify the stock market inefficiency *Physica A Statistical Mechanics & Its Applications* 389(9):1891-1901.

[4] Zhang C, Wang H and Fu R 2014 Automated Detection of Driver Fatigue Based on Entropy and Complexity Measures *IEEE Transactions on Intelligent Transportation Systems* 15(1):168-177.

[5] Cao JW, Wan YX, Tu GY, et al 2013 Information system architecture for smart grid *Chinese Journal of Computers* 36(1):143-167

[6] Kim KD and Kumar P R 2012 Cyber-physical systems:A perspective at the centennial *Proceedings of the IEEE* 100(Special Centennial Issue):1287-1308

[7] Ericsson G N 2010 Cyber security and power system commumication-essential parts of a smart grid infrastructure *IEEE Transactions on Power Delivery* 25(3):1501-1507

[8] Liu XM and Ju J 2013 Analysis of the complexity features for information management system of smart grid *Journal of North China Electric Power University (Social Sciences)* (02):20-24

[9] Liu XM, Tian HY and Qin C 2012 Research on smart grid information management system based on complex scientific management thinking and multi agent system technology *Power System Technology* 36(8):204-208

[10] Song HL, Jin ZX, Li JK, et al 2006 The Three Dimension Metric Model of Management System Complexity for Enterprise *Journal of industrial engineering and engineering management* 20(1):103-108.

IOP Conf. Series: Journal of Physics: Conf. Series **1072** (2018) 012013 doi:10.1088/1742-6596/1072/1/012013

Optimization Strategy for Day-ahead Scheduling of Micro-grid Based on the Cost-effective of User Side

Yang Cui[1], Zifeng Li[1], Wuzhi Zhong[2], Zhiwen Yang[1], Yuting Zhao[1] and Yu Qu[1]

Northeast Electric Power University, Jilin 132012, Jilin Province, China
China Electric Power Research Institute, Haidian District, Beijing 100192, China

nedulzf@163.com

Abstract. Micro-grid, with its important role in coordinating and controlling between distributed generations and the grid, has received widespread attention. Due to the randomness and fluctuation of the new energy output, the access to new energy has adversely affected the economic operation of micro-grid-containing power systems. However, if a reasonable ratio of the new energy is determined when developing the day-ahead scheduling strategy of the micro-grid, it can effectively ensure the economic benefits during the operation. This paper proposes a day-ahead scheduling optimization for micro-grid based on the cost-effective of user side. Various constraints related are taken into consideration, such as charging and discharging margins of storage devices, battery state of charge and generation capacity of the new energy, and the capacity of the electric transmission line. The model is constructed with the goal of minimizing the power supply costs. The prices of time-of-use electricity and the issues of energy exchange between the micro-grid and the external power grid are included as well. As for the method, Cplex is used for its good performance in solving mathematical programming models combined with the operation data of the micro-grid, hence, an optimized day-ahead scheduling of the micro-grid would be generated.

1. Introduction

At present, China's new energy grid-based approach to consumption can be broadly divided into two categories. One is the concentrated delivery, the other one is the local consumption. According to the data from the National Energy Administration, by the end of 2017, the cumulative installed capacity of wind power and solar power in China reached 164 million and 130 million KW. The high proportion of renewable energy access poses new challenges to the safe and stable operation of the power grid. If the new energy is consumed locally, it will be unified with the load and energy storage devices to form a micro-distribution system, which is called "micro-grid". To a certain extent, the micro-grid can reduce the adverse effects of distributed energy access on the grid. Although the micro-grid is less than the large-scale grid, it has several advantages. One of those is its internal application of wind, light, storage for joint power generation. The others include the excellent economy and the resulting environmental benefits. Due to these advantages, micro-grid has been further promoted in the power system. Therefore, the study of scheduling of micro-grid becomes significant.

In [1], it included the wind power in the industrial micro-grid, which suppressed the load fluctuation as well as reducing the power generation cost of the micro-grid. However, it did not consider the complementary cooperation of other new energy sources. In [2], Alternating Direction Method of Multipliers (ADMM) was used to completely decouple the various regions in the micro-grid, hence to economically dispatch the interior of the micro-grid. But it did not take the energy exchange between the micro-grid and the external power grid into account. According to [3], the load model in the micro-grid was carefully classified, in order to minimize the daily operating cost of the system. Based on which, [3] formulated the corresponding scheduling strategy. However, the effect of the transmission capacity of the line on the power exchange was not considered, though the bi-directional power flow between the micro-grid and the external power grid was. Literature [4] fully considered both the power exchange and the role of power market and then proposed a profit-optimal scheduling strategy based on game theory. But the process of energy storage participation was simplified.

In order to give full play to the regulatory capacity of new energy, this paper considers the constraints of energy storage battery's own performance and the transmission capacity of the grid and grid lines. And then, this paper establishes a mathematical model of micro-grid optimal scheduling of micro-grid with the goal of minimizing the average load power supply cost. Using peak-valley electricity prices, micro-grids can charge batteries at the low prices and discharge at the peak ones, hence reduces the costs. Additionally, the optimal ratio of the new energy at different times can be determined, which helps to study the optimal operation strategy of micro-grid under time-sharing electricity price mechanism.

2. Generation of the optimized day-ahead scheduling of the micro-grid

Based on the operation situation of the main network, the whole day is divided into three periods: peak, flat and valley. And the electricity prices are lower in the valley and flat periods and higher in the peak periods[5]. Thus, the micro-grid can use the market to obtain more profit margin, improve the operational efficiency of the micro-grid, and reach a win-win situation between the micro-grid and the main network.

Different scheduling strategies are adopted according to different time periods where the current scheduling moment is located, and different ranges where the battery state of charge is located.

In the cases of full use of renewable energy and abandonment of wind and light, the composition of the load power supply for each period, the total cost of electricity for the whole day, and the load average purchase price are calculated. After a comparative analysis, the cost of optimal conditions for the amount of new energy into the network is determined.

3. Establishment of the optimized day-ahead scheduling of the micro-grid

3.1. The objective function

The objective function is constructed with the goal of minimizing the average load power supply cost, as in (1).

$$\min C = C_{new} + C_{buy} + C_{battery} - W_{sell} \tag{1}$$

The expressions in the above formula:

$$C_{new} = \sum_{t=1}^{T} \frac{24}{T}\left(C_w P_{wplant,t} + C_v P_{vplant,t}\right) \tag{2}$$

$$P_{wplant,t} = P_{wsell,t} + P_{wl,t} + P_{wb,t} \tag{3}$$

$$P_{vplant,t} = P_{vsell,t} + P_{vl,t} + P_{vb,t} \tag{4}$$

Where: C_{new} is the new energy power generation costs; C_w is the cost of wind power generation; C_v is the cost of PV power generation; $P_{wplant,t}$ is the output of wind power at time t; $P_{vplant,t}$ is the output of PV power at time t.

$$C_{buy} = \sum_{t=1}^{T} \frac{24}{T} C_{buy,t} P_{buy,t} \tag{5}$$

$$P_{buy,t} = P_{buyl,t} + P_{buyb,t} \tag{6}$$

Where: $C_{buy,t}$ is the time-of-sale electricity price of the grid; $P_{buy,t}$ is the power purchased by the micro-grid from the grid, $P_{buyl,t}$ is the power obtained from the grid by the load; $P_{buyb,t}$ is the power of the storage battery from the grid.

$$C_{battery} = \sum_{t=1}^{T} \frac{24}{T} C_{dis} P_{dis,t} \tag{7}$$

$$P_{dis,t} = P_{disl,t} + P_{disg,t} \tag{8}$$

Where: $C_{battery}$ is the cost of electricity purchase; C_{dis} is the battery charge to discharge cost; $P_{dis,t}$ is the battery discharge power; $P_{disl,t}$ is the battery power supply to the load; $P_{disg,t}$ is the battery power supply to the grid.

$$C_{sell} = \sum_{t=1}^{T} \frac{24}{T} C_{sell,t} P_{sell,t} \tag{9}$$

$$P_{sell,t} = P_{wsell,t} + P_{vsell,t} + P_{bsell,t} \tag{10}$$

Where: C_{sell} is the electricity sales of micro-grids; $C_{sell,t}$ is the selling power of the micro-grid; $P_{wsell,t}$, $P_{vsell,t}$, $P_{bsell,t}$ are the sales power of the wind power, PV and battery to the grid respectively.

$$Q_{LOAD} = \sum_{t=1}^{T} \frac{24}{T} P_{LOAD,t} \tag{11}$$

Where: Q_{LOAD} is the total daily load; $P_{LOAD,\ t}$ is the real-time load of micro-grid.

3.2. Model constraints

Sections should be numbered with a dot following the number and then separated by a single space:

1） Consider the energy storage battery's own performance

Charge state upper and lower bound:

To prevent the overcharge and the over discharge of the battery, the state of charge of the battery (SOC, State-of-Charge, which is the ratio of remaining battery capacity to battery capacity) should meet the following constraints[6].

$$S_{\min} \le S_t \le S_{\max} \tag{12}$$

Where: $S_{\max}$ is the upper limit of charge state; $S_{\min}$ is the lower limit of charge state; S_t is the state of charge at time t.

Charge and discharge state constraints:

Considering the truth of that in the same time interval, the battery cannot be in charge and discharge at the same time, the battery charge and discharge conditions need to meet the following constraints.

$$X_t \cdot Y_t = 0 \tag{13}$$

Where: X_t, Y_t is a 0,1 variable which representing the charge and discharge status.

Energy state periodic constraints:

Its energy status needs to be equal throughout the scheduling period.

$$S_0 = S_T \tag{14}$$

Where: S_0, S_T is the battery initial and the end of the scheduling cycle SOC state.

Charge and discharge maximum power per unit time constraints:

$$\begin{cases} 0 \le P_{cha,t} \le 0.2 E_b X_t \\ 0 \le P_{dis,t} \le 0.2 E_b X_t \end{cases} \tag{15}$$

Where: $P_{cha,t}$ is the battery charge power; $P_{dis,t}$ is the battery discharge power; E_b is the battery capacity.

Charge and discharge times constraints:

$$\begin{cases} \sum_{t=1}^{T} |X_{t+1} - X_t| \le N_1 \\ \sum_{t=1}^{T} |Y_{t+1} - Y_t| \le N_2 \end{cases} \tag{16}$$

Where: N_1, N_2 are the number of battery charge and discharge within a day limit respectively.
Consider storage battery charge balance and discharge balance:

$$\begin{cases} P_{cha,t} = P_{buyb,t} + P_{wb,t} + P_{vb,t} \\ P_{dis,t} = P_{disl,t} + P_{disg,t} \end{cases} \tag{17}$$

Where: $P_{cha,t}$ is the battery charging power.

2) Consider the tie line transmission capacity

$$\begin{cases} P_{buy,t} \le P_{\max} \\ P_{sell,t} \le P_{\max} \end{cases} \tag{18}$$

Where: $P_{\max}$ is the contact line transmission capacity.

3) Consider the balance between loads

$$P_{wl,t} + P_{vl,t} + P_{disl,t} + P_{buyl,t} = P_{LOAD,t} \tag{19}$$

4) New energy consumption constraints

When new energy is completely consumed:

$$\begin{cases} P_{w,t} = P_{wsell,t} + P_{wl,t} + P_{wb,t} \\ P_{v,t} = P_{vsell,t} + P_{vl,t} + P_{vb,t} \end{cases} \tag{20}$$

When new energy can be not completely consumed:

$$\begin{cases} P_{w,t} \ge P_{wplan,t} = P_{wsell,t} + P_{wl,t} + P_{wb,t} \\ P_{v,t} \ge P_{vplan,t} = P_{vsell,t} + P_{vl,t} + P_{vb,t} \end{cases} \tag{21}$$

4. Case study

4.1. Parameter settings

To schedule, 24 hours a day will be divided into 96 time periods at 15-minute intervals in this paper. It is known that wind power installed capacity is 250kW, power generation cost is 0.52 yuan / kWh, PV installed capacity is 150kW, and power generation cost is 0.75 yuan / kWh. Disregarding the battery loss, the rated capacity of the battery is 300 kWh, the operating range of the battery SOC is [0.3, 0.95], the initial SOC value is 0.4, the cost from charging to discharging is 0.2 yuan / kWh, the daily number of charge and discharge restrictions is 8 times, and the tie line transmission capacity is 150kW. Amongst them, the prices of each period on the grid side are shown in Table 1.

Table 1. The prices of each period on the grid side.

Time(h)	0:00-7:00	7:00-10:00	10:00-15:00	15:00-18:00	18:00-21:00	21:00-24:00
Sales price (yuan)	0.22	0.42	0.65	0.42	0.65	0.42
Purchase price(yuan)	0.25	0.53	0.82	0.53	0.82	0.53

4.2. Model solving

The economic operation of micro-grid is a multi-dimensional, dynamic optimization problem with complex equations and inequality constraints[7]. This paper uses Cplex software to solve the model.
From the optimization results shown in Fig. 1, due to the full utilization of renewable energy, the PV power supply is not started at the initial stage, and all the load power supply is provided by the wind power. The battery output curve shows that when the battery starts to supply power to the load, the total amount of new energy output generated is less than the load. At this time, wind power, PV, and battery start to supply power to the load. When involving the battery, power supply is still unable to meet the needs, the micro-grid began to purchase electricity from the grid side. At the 60th scheduling

moment, after 15:00, the output of wind power and PV power decreases, and the load start to increase. Only purchasing electricity from the grid can maintain the load balance.

As can be seen from Fig. 2, the battery is already fully charged before the 28th scheduling time, which is before 7:00. At the 56th scheduling time, around 14:00, the storage capacity begins to drop, indicating that the discharge process has started. Combined with Figure 1, the battery begins to supply power to the load in this period. At the 64th scheduling time, at 16:00, it starts to continue charging. From the time-of-use price, it is clear that during this period, the electricity purchase price is at flat and valley moment. And according to the load and the predicted power of renewable energy, in the next period, there is not enough renewable energy to supply the load. What's more, the electricity purchase price is at the peak. Therefore, the battery chooses to charge in the flat and valley period, at 16:00, and to discharge in the peak period. This reflects the battery's role of "energy real-time shift", thereby reducing the cost of purchasing electricity from the external power grid and improving the economic benefits of the micro-grid itself.

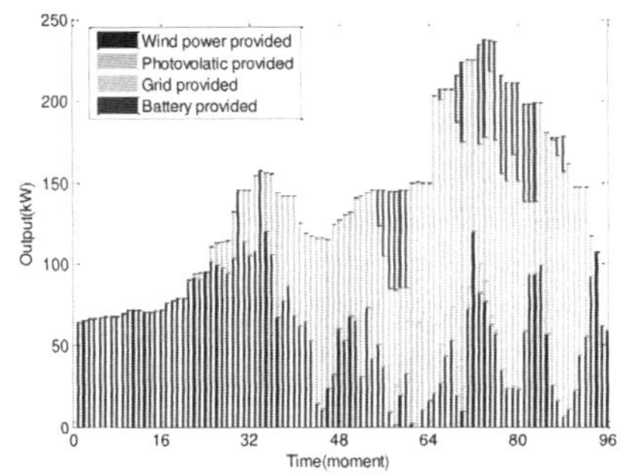

Figure 1. Load power supply composition when the renewable energy completely absorbed.

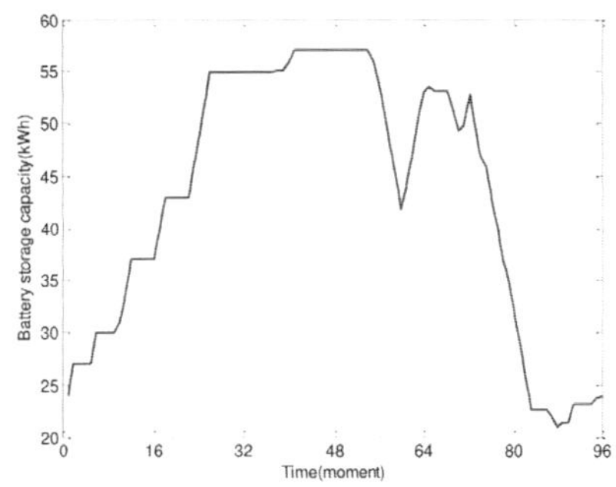

Figure 2. Battery storage capacity when the renewable energy completely absorbed.

Figure 3 is the composition of each period of the power supply when allowing the abandonment of wind and light. PV output is smaller but relatively concentrated in the period when electricity price is higher. The dispatching scheme purchases a large amount of electricity from the grid, and the battery supply to the load only appears when the exchange power of the micro-grid and the grid exceeds 150 kW or the purchase price of electricity is high.

When the wind and light conditions are allowed to abandon, the battery storage capacity is shown in Fig.4. The battery can store PV, wind power and electricity purchased from the grid. It can also choose the storage object after comparing these three prices. When the electricity purchase price is high or when the exchange power of the micro-grid and the grid exceeds 150kW, it meets the precondition of power switching constraint and achieves the goal of minimizing the load power supply cost

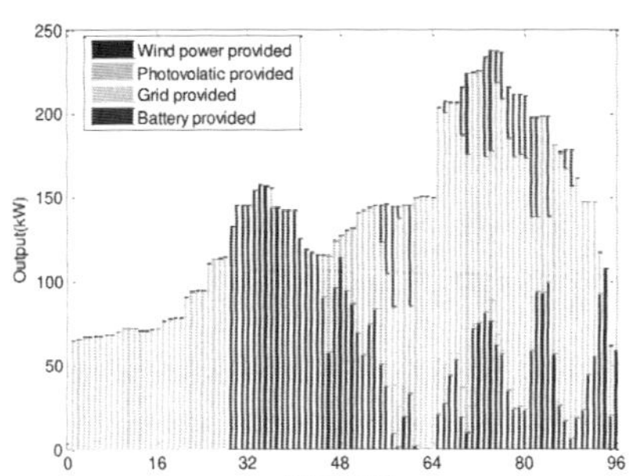

Figure 3. Load power supply composition when new energy not completely consumed.

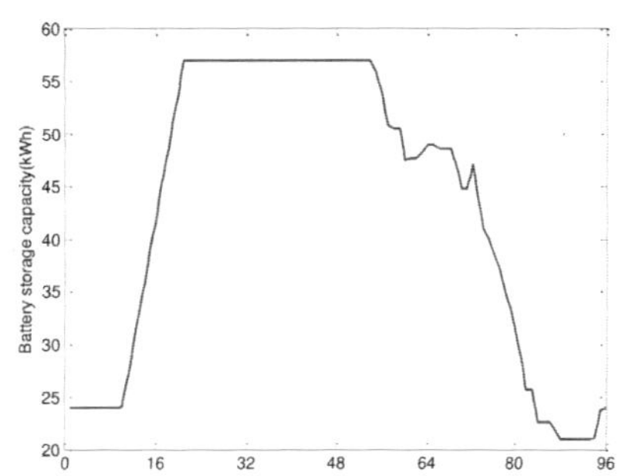

Figure 4. Battery storage capacity when new energy can be not completely consumed.

Calculated in full use of new energy sources, the total cost of electricity for the whole day is 2212.46yuan. The average purchase price of electricity is 0.625yuan/kWh. Under the condition of abandoning the wind and the light, the average purchase price of the load is 0.4837 yuan / kWh, and the total price of electricity supply for the day is 1735.49 yuan. The new energy utilization under the two schemes is shown in Figure 5 and Figure 6. The amount of abandoned wind power is 1498.2kWh, and the total wind power generation is 2783.1kWh, with a wind abandonment rate of 53.8%. The amount of abandoned PV power is 408.54kWh, and the total amount of PV power generation is 629.145kWh, with a light abandonment rate of 64.94%.

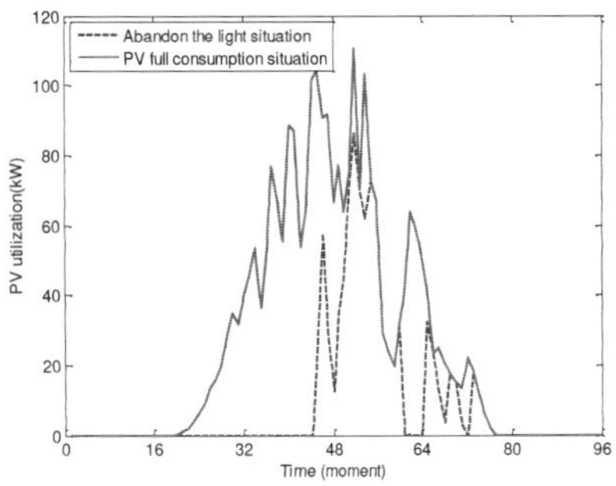

Figure 5. Two scenarios of PV utilization.

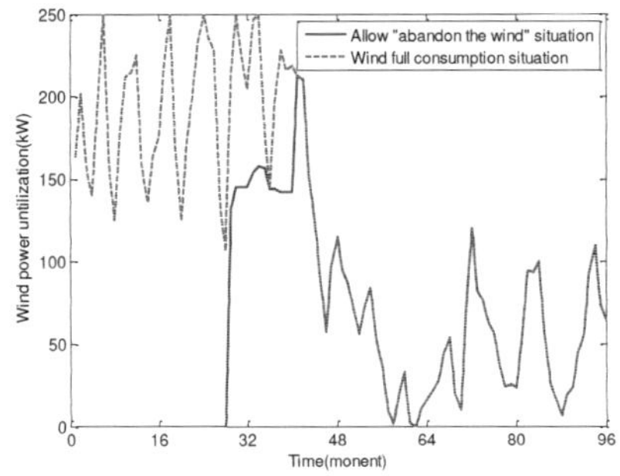

Figure 6. Two scenarios of wind utilization.

5. Conclusion

In this paper, by considering the performance of the energy storage battery and the related constraints of the transmission capacity between the micro-grid and the grid, a mathematic model of optimized operation of micro-grid is established with the goal of minimizing the average load power supply cost. Meanwhile, the optimal ratio of new energy at different times is determined, and a micro-grid optimization network scheduling program is developed, with the purpose of optimizing the cost of power supply. Moreover, the following conclusions are drawn:

Through the example analysis, it can be seen that compared with the total energy consumption of new energy sources, the micro-grid wind abandonment rate is 53.8%, and the light abandonment rate is 64.94% under the condition of the wind and the light abandonments. However, the power supply cost of the whole day is reduced by 476.97 yuan. The average unit price of load decreases from 0.625 yuan / kWh to 0.4837 yuan / kWh, which is a decrease of 22.6%. Although new energy is not fully utilized, the economy of the system is improved.

Above all, improving the economy during the operation of micro-grid does not mean utilizing as much new energy as possible. The new energy generation capacity, the time-of-use electricity prices, the energy storage battery performance should be involved. In addition, store when the main grid electricity price is underestimated and sell when the price is at the peak. After considering all these conditions, an appropriate scheduling strategy will be developed.

6. References

[1] Hao D and Feng G 2015 *Automation of Electric Power Systems* **39** 160
[2] Yangcong O and Mingbo L 2017 *Automation of Electric Power Systems* **32** 134
[3] Qingshan X, Aidong Z and Kai W 2016 *Power System Technology* **40** 1657
[4] Chunxia D, Xingbei J and Heng L 2016 *Power System Technology* **40** 579
[5] Rouhani A and Joorabian M 2017 *International Conference on Recent Advances and Innovations in Engineering* 237
[6] Basu A K, Bhattacharya A and Chowdhury S 2012 *IEEE Trans. Power Systems* **27** 30
[7] Majidi M, Nojavan S and Esfetanaj N 2017 *Solar Energy* **14** 79

IOP Conf. Series: Journal of Physics: Conf. Series **1072** (2018) 012014 doi:10.1088/1742-6596/1072/1/012014

Coordinated Wind Power Accommodating Dispatch Model Based on Combined Electricity and Heat System

Yan Zhuang[1], Yu Qu[1], Wuzhi Zhong[2], Gangui Yan[1]

[1]School of Electrical Engineering, Northeast Electric Power University, Jilin 132012, Jilin Province, China

[2]China Electric Power Research Institute, Haidian District, Beijing 100192, China

zhuangyanjyjy@163.com

Abstract. To solve the problem of wind power accommodation, this paper proposed a coordinated wind power accommodating dispatch model based on combined electricity and heat system with the perspective of decoupling thermoelectric coupling constraints and enhancing the adjustment capacity of the power system. Firstly, this paper introduced the working principles of wind power characteristics and combined heat and power (CHP), the basic operating principles of the consumption of wind power curtailment using thermal storage. Secondly, the coordinated wind power accommodation dispatching model was established due to the regulations mentioned above. Thirdly, the main problem was divided into two sub problems by Benders decomposition, which are power and thermal dispatching, and then the optimum solutions which meet the constraints of each sub problem was solved using Interior point. Finally, the whole dispatching planning of CHP was obtained. Case study verified that, the proposed model is effective.

1. Introduction

Energy and environmental crisis are the major challenges facing the world today. Taking the three major fossil fuels of coal, oil and natural gas as examples, according to the proved reserves, the ratio of production to storage is 153, 50.6 and 52.5 years, respectively[1]. The development and utilization of renewable energy is a realistic choice to ensure the sustainable use of energy. The 'Energy Development Strategy Action Plan(2014-2020)'[2] officially released by the State Council pointed out that it is necessary to vigorously develop renewable energy sources, in accordance with the output and the local utilization of both centralized and distributed development principle, accelerate the development of renewable energy.

As an effective way to deal with the energy crisis, wind power has been rapidly increasing in the world in recent years. The development of wind power in our country started relatively late but developed rapidly. In 2016, the installed capacity of wind power increased by 19.3 million kW and the total installed capacity reached 149 million kW, accounting for 9% of the total installed capacity of power generation,also the electricity of wind power integration reached 241.0 billion kwh. However, the wind power curtailment problem is still very serious. The total amount of wind power curtailment of China exceeded 49.7 billion kwh in 2016. The national average wind power curtailment rate was 17.1%. Among them, the rate of the 'Three North regions' ranked first in the country, the more serious areas are Gansu, Xinjiang and Jilin, the wind power curtailment rate reached 43%, 38%, and 30%[3].

The reason is that on the one hand, the wind power resources are abundant in the 'Three North regions', but the energy consumption is insufficient, so it is difficult to achieve large-scale wind power by the local consumption while being restricted by transmission channels of the power grids[4]. On the other hand, there is a natural contradiction between the stochastic volatility of wind power itself and the safe and stable operation of power grid. In order to meet the demand of heat load in winter, 'the load level of CHP is restricted by the minimum heat demand' production mode further reduces the space for wind power and exacerbates the phenomenon of wind power curtailment.

Energy storage technology has received extensive attention in recent years because of its good wind power complementarity[5]-[6]. With the development of electric boiler heating technology, a system of combined heating by electric boilers and CHP units has begun to emerge in foreign countries. By using electric boiler heating to increase the load of electricity during the trough of low load, the space for wind power access has significantly increased.

Although the development of these technologies has effectively alleviated the contradiction between wind and heat, however, most of the current studies separately consider the role of heat storage device and electric boiler for disposal of abandoned wind. Taking into account the close combination of future power system will be with the thermal system, it is very urgent and necessary to study how to coordinate the work of each unit in the energy system such as electric heating system to reduce the waste of energy loss and obtain the best economy.

2. Working principle

2.1. Wind power characteristics

Both wind power and conventional power supply convert primary energy into secondary energy. Conventional power supplies are based on natural water or fossil fuels as primary energy sources. These energy sources can be treated in a stable, sustainable and controllable manner.The energy source of wind power is wind energy in nature. Affected by factors such as season, climate and topography, wind energy can not artificially control the variation of wind energy even though some predictions of wind randomness can be made.

After the wind power is connected to the grid, due to its randomness, fluctuation, intermittency and anti peak shaving characteristics, it will affect the state of supply and demand balance that the power system is completely controllable, dispatching and tracking the load forecast before the wind power is not connected to the grid. Therefore, the power grid may overestimate the risk caused by wind power fluctuations, thereby limiting the access scale of wind power, so as to ensure the safe and stable operation of the grid.

In summary, in order to improve the ability of the system's wind power accommodation and solve the contradiction between wind power and heating units. Therefore, it is necessary to analyze the working principle of CHP units, which is now the main heating unit in the cogeneration system.

2.2. Working principle of the traditional CHP unit and CHP unit with thermal storage

The coupling relation between the generating power P_{el} and the heating power P_h of the CHP unit is the 'electric heating characteristic'. The common CHP units can be divided into four types which are according to different electric heating characteristics[7]. This paper takes the most common steam extraction CHP unit as an example to analyze its electrothermal characteristics as shown in figure 1.

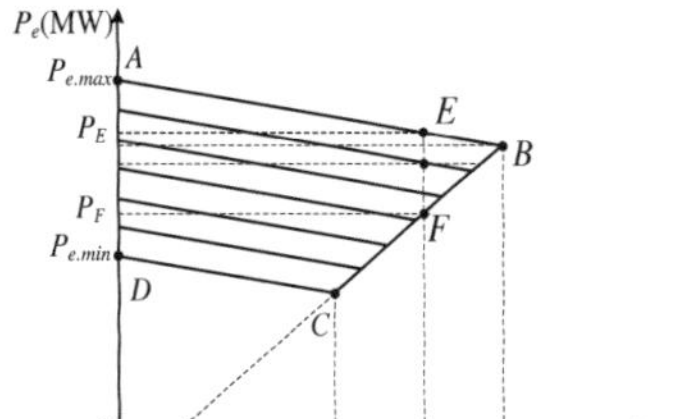

Figure 1. Heat-electricity characteristic of conventional CHP unit.

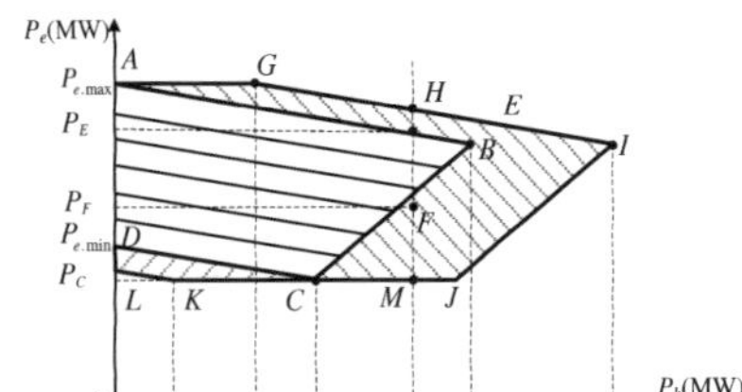

Figure 2. Heat-electricity characteristic of CHP unit with thermal storage.

As can be seen from the operation interval shown in figure 1, the power generation has a certain degree of adjust ability under a given heat load, such as the heating power P_h. Power generation can be adjusted between P_E~P_F, but the larger the heating power, the smaller the range of electric power can be adjusted. This is because, at a given extraction volume, such units can adjust the power output of the entire turbine by adjusting the amount of steam generated by condensing steam, but the larger the extraction volume, the smaller the proportion of condensing steam that can be adjusted. So the smaller the scope of regulation, abatement wind power capacity is very limited.

The traditional CHP unit in the configuration of the thermal storage device, the electric characteristics will be greatly changed, as shown in figure 2. Since the thermal storage device can perform heat storage and heat release operations, assuming that the heat release and storage powers are $P_{h.fmax}$ and $P_{h.cmax}$, the electrically feasible operating field AGIJKL containing the thermal storage CHP unit is actually the superimposed region formed by the translation of the original feasible operating domain ABCD to the right of the horizontal axis $P_{h.fmax}$ and to the left of the translation of $P_{h.cmax}$.

At the same heating power P_h, the power generation of the steam turbine can be allowed to be adjusted between P_M and P_H, and the resulting steam turbine is under-heated or the remaining part of the heating is compensated by the thermal storage device for heating or heat storage to maintain a stable supply of heat load, thereby enhancing the peak capacity of the unit.

2.3. The operation principle of wind power accommodation with thermal storage

Thermal storage is usually built on the heat source side of the heating system and connected between the thermal power plant and the heating network. At present, the main heating units in China are large-scale coal-fired condensing bleeder units, and the start-up and shutdown costs of such units are very high. Therefore, after the heat storage device is configured, peak load can also be adjusted by using the low-load operation mode. The operation principle of wind power accommodation with thermal storage is: during daytime electric load waist load, increase steam turbine inlet steam power and steam extraction power for thermal storage; at electric load low troughs, while reducing the power generation and heating power for peak load down, the heating part of the supplement by the thermal storage.

2.4. Mechanism of wind power accommodation of the coordinated heating system

By arranging the thermal storage device on the side of CHP unit, not only the purpose of decoupling the thermoelectric coupling characteristics can be achieved, but also the power system can be closely connected with the heating system to form an integrated electric heating system to improve the power system optimal allocation ability and enhance the ability of consuming the abandoned wind. However, only the configuration of heat storage will still appear abandoned wind phenomenon, if the load side accesses electric boiler and CHP unit coordinated heating, the wind power accommodation space of the power grid can be further expanded. The electric boiler is used to increase the electricity consumption when the load is low, and the role of heat storage is to reduce the power output of CHP unit and improve the space of wind power network.The structure of the integrated system containing CHP units with thermal storage and electric boilers is shown in figure 3.

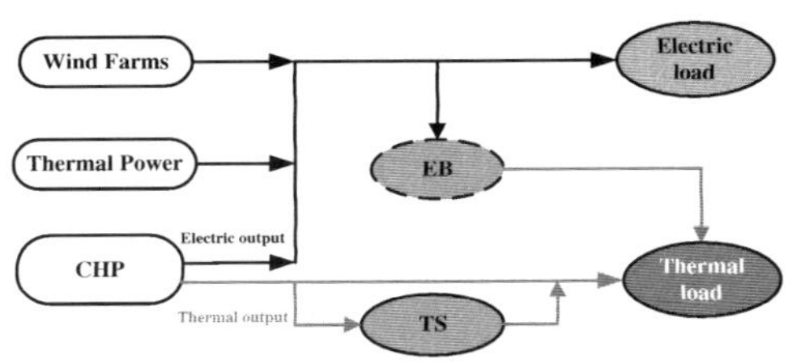

Figure 3. Structure diagram of combined heat and power system.

Integrated system access to electric boiler, on the one hand, CHP unit's range will be more expanded through the electric boiler; on the other hand, electric boiler access while increasing the

IOP Conf. Series: Journal of Physics: Conf. Series **1072** (2018) 012014 doi:10.1088/1742-6596/1072/1/012014

electric load, and the role of the thermal storage device at the same time makes the power grid abatement wind power greatly enhance the ability to alleviate the problem of wind power curtailment.

3. Coordinated wind power accommodating dispatch model based on combined electricity and heat system

$$\begin{cases} P_{w.h.t} = P_{w0.t} - \Delta P_{w.h.t} \\ \Delta P_{w.ech} = \Delta P_{w.ch} + \Delta P_{EB} \\ P_{w.h.t} = \Delta P_{w.ech} \end{cases} \tag{1}$$

Where: $P_{w.h.t}$ is the abandoned wind power of the system only containing CHP unit at time t, $P_{w0.t}$ is the abandoned wind power of the system only containing CHP unit without heating at time t, $\Delta P_{w.h.t}$ is the space of $P_{w.h.t}$(maybe negative), $\Delta P_{w.ech}$ represents the abandoned wind space which is produced by CHP with thermal storage and electric boiler in coordinated heating, $\Delta P_{w.ch}$ is the abandoned wind space of CHP with thermal storage, ΔP_{EB} is the abandoned wind space of electric boiler.

When the total heating load P_{hz} is between point C and J in figure 2, for a system with a single electric boiler and CHP unit, the heat elimination combined heat and power unit can generate heat when the heat is removed. The space is expressed as a function of the heating power $P_{H.e}$ of the electric boiler.

$$\Delta P_{w.ch} = \begin{cases} c_m(P_{hz} - P_{H.e}) - (c_v + c_m)P_{hT,\max} + P_{el,\max} - P_{el,\min} + c_v P_{h,med}; \\ P_{H.e} \in [0, P_{hz} - P_{h,med}] \\ c_v \cdot P_{H.e} + c_v(P_{h,med} - P_{hz}); P_{H.e} \in (P_{hz} - P_{h,med}, P_{hz} - P_{h,med} + P_{h.c\max}] \\ c_v \cdot P_{h.c\max}; P_{H.e} \in (P_{hz} - P_{h,med} + P_{h.c\max}, P_{hz}] \end{cases} \tag{2}$$

Where: c_m is the elastic coefficient of electric and thermal power at back pressure, c_v is the reduction of power generation by multiple extraction of unit heat with constant intake of steam, $P_{H.e}$ is heating power of electric boiler, $P_{h.cmax}$ is the maximum heat storage power of the thermal storage device, $P_{h,med}$ is the turbine heating power when the generating power is the minimum, $P_{hT,max}$ is the maximum heating power of the unit, $P_{el,min}$, $P_{el,max}$ are the minimum, maximum power output of the unit under pure condensing conditions, respectively.

3.1. The objective function of scheduling model

Economic dispatch problems with wind power systems usually take the minimum system cost as the dispatch target. In order to test the effect of the heat storage device and the electric boiler to dissipate the wind power, the cost of abandoning the wind is added to the cost, so the objective function is:

$$\min\left(F_1(P_i) + F_2(P_{erl.i}, P_{h.i}, P_{cr.i}) + \lambda \cdot P_{w.qf}\right) \tag{3}$$

Where: P_i is the electric output of the conventional unit *i*, $P_{erl.i}$ is the electric output of the CHP unit *i* with thermal storage, $P_{h.i}$ is the total heating power of the CHP unit *i* with thermal storage, $P_{cr.i}$ is the heat release and storage powers of the CHP unit *i* with thermal storage, $P_{w.qf}$ is the system's abandoned wind power, λ is the cost coefficient of abandoned wind.

(1)The costs of thermal power generation F_1

Wind power integration has an important influence on the power unit output size, the unit start stop plan and the start-up and shutdown of the unit during operation[8]:

$$F_1 = f_1(P_i) + f_2(u_i) \tag{4}$$

Where: f_1 indicates the operating costs of the thermal power unit, f_2 indicates thermal power start-stop plan and start-stop peak-regulation costs, P_i, u_i are the power generated and operating status of unit *i*.

$$f_1 = \sum_{t=1}^{T}\sum_{i=1}^{N}\left(a_i P_{i.t}^2 + b_i P_{i.t} + c_i\right) \tag{5}$$

$$f_2 = \sum_{t=1}^{T}\sum_{i=1}^{N}\left[u_{i.t}(1 - u_{i.t-1})S_i\right] \tag{6}$$

Where: a_i, b_i, c_i are the quadratic fitting coefficient of the conventional unit operating costs, $P_{i.t}$ indicates the power output at *t* moment for conventional unit *i*, $u_{i.t}$ indicates the status of start or stop at *t* moment for unit *i*, $u_{i.t}$=1 indicates run， $u_{i.t}$=0 indicates stop, S_i is the start-up cost of unit *i*.

(2)The costs of CHP power generation F_2

$$F_2 = \sum_{t=1}^{T}\sum_{i=1}^{R}\left\{a_i\left[P_{erl.i}^t + c_v\left(P_{h.i}^t + P_{cr.i}^t\right)\right]^2 + b_i\left[P_{erl.i}^t + c_v\left(P_{h.i}^t + P_{cr.i}^t\right)\right] + c_i\right\} \tag{7}$$

Where: a_i, b_i, c_i are coefficient of operating costs for CHP units, $P^t_{erl.i}$, $P^t_{h.i}$, $P^t_{cr.i}$ are respectively the power output, the total heating power, the heat storage device storage and the heat release power of the CHP unit i at the t moment.

3.2. *Running constraints*

The constraints on the scheduling model of dissipate the wind power for coordinated heating include the following two parts.

(1) Power system constraints

Load balancing constraints:

$$\sum_{i=1}^{N+R} P_{el.i}^t + \sum_{j=1}^{K} P_{wf.j}^t - P_{L.t} - P_{EB}^t - P_{w.qf}^t = 0 \tag{8}$$

Where: $P_{el.i}^t$ represents the power generation of the conventional thermal power or CHP unit i at the t moment, $P_{wf.j}^t$ indicates the predicted output of the wind turbine j at the t moment, P_{EB}^t is the heating power consumed that connected to the electricity boiler at the t moment, $P_{w.qf}^t$ indicates the abandoned wind power at the t moment.

Unit output constraints：

$$P_{el,\min i} \le P_{el,i}^t \le P_{el,\max i} \tag{9}$$

Where: $P_{el,\min i}$ is the minimum generation power of a conventional thermal power or CHP unit i, $P_{el,\max i}$ represents the maximum power generation of a conventional thermal or CHP unit i. According to the above, the upper and lower limits of the power output for the CHP unit are a function of the heating power, while the conventional power is a fixed value.

The ramp of unit constraints：

$$-P_{d.i}^t \le P_i^t - P_i^{t-1} \le P_{u.i}^t \tag{10}$$

Where: $P_{d.i}^t$, $P_{u.i}^t$ represent the ramp rate of upward and downward for unit i at the t moment, P_i^t is the power generation of unit i at the t moment.

(2) Thermal system constraints

Operation constraints for thermal storage：

$$\begin{cases} S_i^t - S_i^{t-1} = P_{cr.i}^t \le P_{h.c\max.i} \\ S_i^{t-1} - S_i^t = P_{cr.i}^t \le P_{h.f\max.i} \\ S_i^t \le S_{i.\max} \\ \sum_{t=1}^{T} P_{cr.i}^t = 0 \end{cases} \tag{11}$$

Where: S_i^t is the heat storage capacity of the thermal storage device i at the t moment, $P_{h.c\max.i}$ $P_{h.f\max.i}$ respectively represent the maximum heat storage and release power of the thermal storage device i, $S_{i.max}$ is the heat storage capacity of the thermal storage device i, $P_{cr.i}^t$ is the heat storage and release power of the thermal storage device i at the t moment.

Heating balance constraints：

$$P_{H.e}^t + \sum_{i=1}^{R} P_{h.i}^t = P_{hz}^t \tag{12}$$

Where: $P_{H.e}^t$ is the heating power of electric boiler at the t moment, $P_{h.i}^t$ is the total heating power of the CHP unit with heat storage at the t moment, P_{hz}^t is the heating load at the t moment.

3.3. *Solve the model*

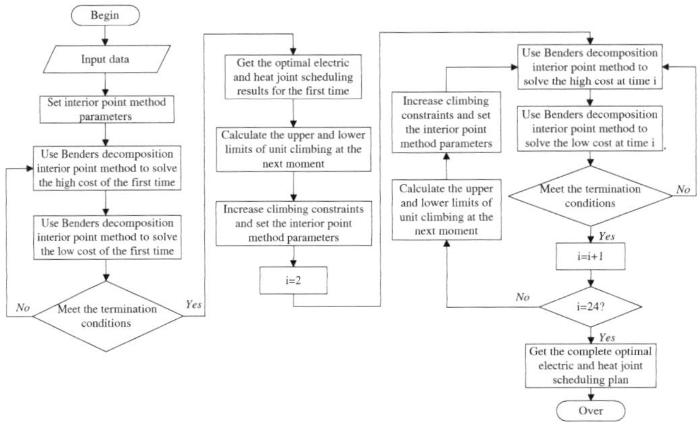

Figure 4. Schedule flow chart of Benders decomposition interior point method.

4. Examples

4.1. Example conditions

The cost coefficient of abandoned wind λ is 100 USD/MW·h[9]. The example is based on the literature [10]. The electric heating system consists of four units, of which unit #1 is a conventional thermal power unit, and unit #2 and #3 are CHP units, the electric heating characteristic curve is shown in figure 5, unit #4 is a pure heating unit, and the system cost function is:

$$\mathrm{COST} = \mathrm{cost1} + \mathrm{cost2} + \mathrm{cost3} + \mathrm{cost4} \qquad unit: dollar \tag{13}$$

$$\begin{cases} \mathrm{cost1} = 50P_1 \\ \mathrm{cost2} = 2650 + 14.5P_2 + 0.0345P_2^2 + 4.2h_2 + 0.03h_2^2 + 0.031P_2h_2 \\ \mathrm{cost3} = 1250 + 36P_3 + 0.0435P_3^2 + 0.6h_3 + 0.027h_3^2 + 0.011P_3h_3 \\ \mathrm{cost4} = 23.4h_4 \end{cases} \tag{14}$$

Constraints include load balancing constraints, heat balance constraints, the ramp of unit constraints and the constraints of the unit output range. In order to find the nearest node from any coordinate distance, the initial value of low cost is taken as $-\infty$, and then replace it constantly until the nearest node is found. ε is taken as 0.0001, when the $|C_{up}-C_{down}|\leq\varepsilon$, that is, when the difference between low cost and high cost is less than ε, the iteration stops.

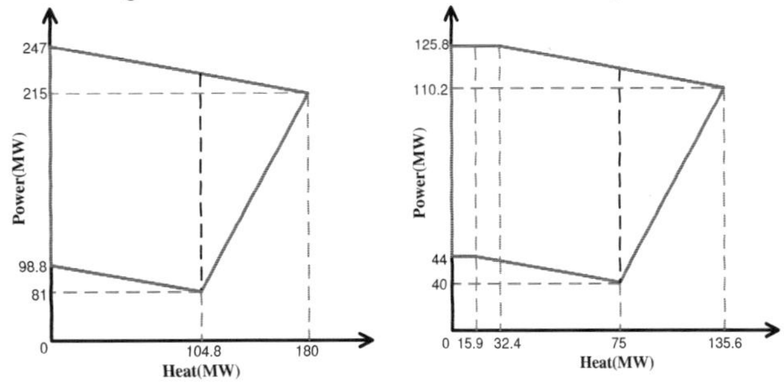

(a)unit #2 (b) unit #3

Figure 5. Heat-electricity characteristic of unit #2, #3.

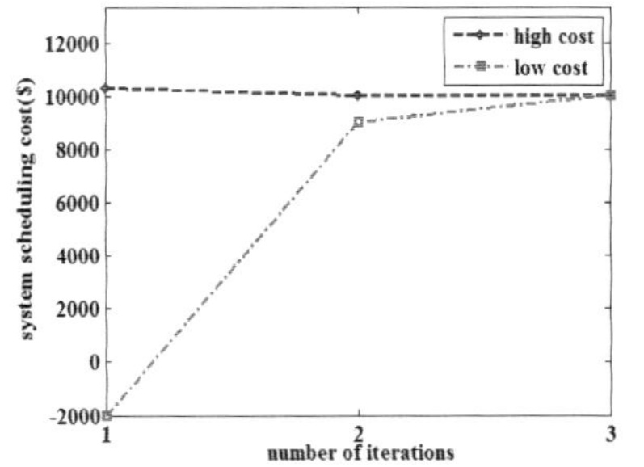

Figure 6. The iteration curve of Benders for the first time.

4.2. Example results

Taking the first moment as an example, the iterative curve of Benders decomposition interior point method is shown in figure 6. The best power output plan for each unit is shown in figure 7 and the best heat output plan is shown in figure 8.

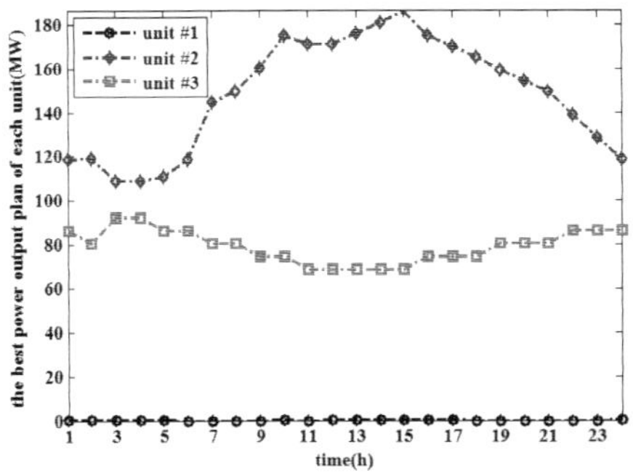

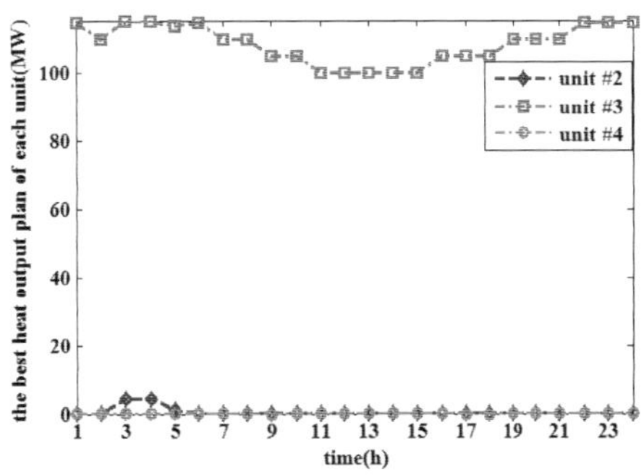

Figure 7. The best power output plan of each unit. **Figure 8.** The best heat output plan of each unit.

According to the result of the scheduling, it is found that #1 conventional thermal power unit is not powered and #2 and #3 CHP units participate in the power supply arrangement as the best power output plan, #4 conventional thermal power unit does not heat, #2 and #3 CHP units to participate in the heating arrangement for the best heat output plan, with this scheduling arrangement, the system can achieve the best economy. The scheduling cost of each period of the system is shown in figure 9.

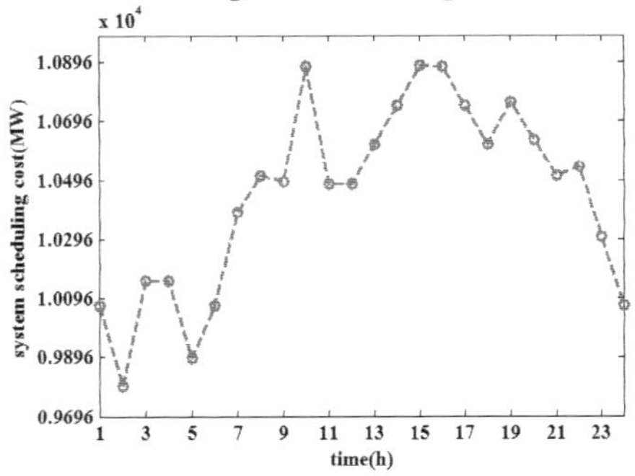

Figure 9. System scheduling costs of each period.

The total scheduling cost of the system is 250718.1767 dollars, at this time, the system does not abandon the wind, and the scheduling cost is reduced by 9.02% than that of not adding the thermal storage device and the electric boiler.

5. Conclusion

(1)By configuring the heat storage unit on the CHP unit side, not only to achieve the purpose of decoupling thermoelectric coupling characteristics, but also to connect the power system with the thermodynamic system. Thus forming a comprehensive electric heating system to improve the power system optimal allocation ability. However, it is still possible to abandon the wind only by configuring heat storage. At this time, if the load side access to electric boilers and CHP units coordinated heating. On the one hand, through the coordinated heating of the electric boiler and the heat storage device, the range of the electric power adjustment of the combined heat and power unit will be further expanded; on the other hand, electric boiler access while increasing the electrical load, with the heat storage device making the ability of eliminating wind power greatly enhance for the power grid at the same time, which can alleviate the problem of abandoning the wind.

(2)In this paper, Benders decomposition of interior point method can effectively solve the model.

(3)The example shows that the coordination scheduling model proposed in this paper is correct and effective. According to the scheduling results found that the scheduling of the pure heating unit is not heating and the pure power unit is not powered, the system can achieve the best economy.

6. References

[1] Energy strategy research group of the Chinese Academy of Sciences 2009 *China's road map for the development of energy science and technology to 2050* (Beijing: *Chinese Science Press*)

[2] *'Strategic action plan for energy development (2014-2020)'*. http://news.dichan.sina.com.cn

[3] National Energy Administration 2017-01-26 *'Wind power grid operation of 2016'*. http://www.nea.gov.cn/2017-01/26/c_136014615.htm
[4] Mu Gang, Cui Yang, et al, 2013 *Automation of Electric Power Systems* **37** 24-29
[5] Yuan Xiaoming, Cheng Shijie, et al, 2013 *Automation of Electric Power Systems* **37** 14-18
[6] Yan Gangui, Liu Jia, et al, 2013 *Proceedings of the CSEE* **33** 45-52
[7] Provas Kumar Roy, Chandan Paul, et al, 2014 *Electrical Power and Energy Systems* **54** 392-403
[8] Cui Yang, Zhang Chao, et al, 2014 *Power System Technology* **38** 2174-79
[9] Ren Jianwen, Xu Yingqiang, et al, 2018 *Power System Technology* **42** 1079-86
[10] Hamid Reza Abdolmohammadi, Ahad Kazemi, 2013 *Energy Conversion and Management* **71** 21-31

Investigation of MMC-HVDC Operation Performance Based on Improved Control Strategy

Li Sun[1*], Feng Hu[1], Sitong Liu[2]

[1] School of Electrical Engineering, Northeast Electric Power University, Jilin, Jilin 132012, China
[2] Guodian Science and Technology Research Institute, Shenyang, Liaoning 110102, China

*Li Sun: sunlinedu@163.com
Feng Hu: 1506405321@qq.com

Abstract. Modular multilevel converter (MMC), as a new type of topology, has aroused widespread concern in the research on HVDC system. However, there are some problems such as low anti-interference performance and output instability in their internal control systems. Based on these problems, this paper improves pole control on original control strategy. The improved control strategy not only can achieve active and reactive power independent control as well as stability control of the DC voltage, but also can reduce the oscillation amplitude and cut down the recovery time, as a consequence, the transient and steady-state performance are improved efficaciously. This paper establishes MMC-HVDC two AC systems and conducts the contrastive simulation analyses between the improved control strategies and original strategies in the MATLAB/Simulink. The results verify the validity and effectiveness of the theory.

1. Introduction

With the rapid development of HVDC transmission technology, the high voltage direct currentbased on voltage source converter (VSC-HVDC) has been widely applied to large city load centers, offshore drilling platforms, island power supply, new energy and other HVDC transmission projects because of its independent control of active and reactive power, two-way transmission of energy, adaptability in various transmission occasions, no commutation failure and many other advantages [1-4].

Among them, MMC acts as a new topology of VSC, which has great prospects for development due to its low switching frequency, high waveform quality and scalability [5-6]. For these reason, MMC becomes an important object for the future development of HVDC transmission projects. The MMC is made up of several sub modules controlled by PWM control signals. Any changes of sub module operation parameters will affect the output of the entire waveform. Therefore, compared with VSC-HVDC, MMC control strategy under steady state operation requires additional sub module capacitor voltage balancing control and circulating current suppression control [7], besides the basic control strategy deduced from the AC side mathematical model. The complexity and diversity of control strategy make the control requirement of MMC-HVDC system very strict for all links. A weakness in a link control will inevitably affect other links.

The VSC-HVDC control strategy mainly contains the current inner loop and voltage outer loop control system in [8]. For the two level, three level VSC-HVDC, due to the additional AC side filter with large capacity and the DC side capacitor, the output stability is greatly enhanced, so the traditional control strategy can satisfy the requirements of normal operation. However, the original VSC control strategy cannot be perfectly applied to the MMC-HVDC system with the increase of the voltage balancing control in the structure. Therefore, the control strategy must be improved to effectively suppress the abnormal conditions and small disturbance waves in the MMC-HVDC system, and reduce the amplitude of the fault oscillation as well as the recovery time after the failure.

According to these above-mentioned requirements for MMC controller, new control elements are added in system control to limit oscillation amplitude and reduce disturbance recovery time, so as to improve the stability and anti-interference of the control system. The MMC-HVDC two AC system based on improved control strategy is built in MATLAB/Simulink environment to simulate steady and transient stage, analyze the operation performance before and after the improvement. The results verify the correctness and validity of the improved control strategy.

2. MMC-HVDC System Control Strategy

At present, the general outer loop controller control is relatively simple, as shown in Fig. 1, P_{ref}, Q_{ref}, U_{dcref} are respectively the reference active power, reactive power and DC voltage value; P_{meas}, Q_{meas}, U_{dcmeas} are respectively the measurement value of active power, reactive power and DC voltage, i_{dref}, i_{qref} are respectively dq reference value of the AC current.

By comparing the actual measured value P_{meas}, Q_{meas}, U_{dcmeas} with the reference value P_{ref}, Q_{ref}, U_{dcref}, the unbalance quantity can be obtained, and then convert it into an inner loop current controller input quantity i_{dref}, i_{qref} through the PI controller. The control type belongs to open loop control.

Although the traditional outer loop controller shortens the control response time to a certain extent, its structure is not able to control the abnormal working condition or disturbance accurately, so that the reliability and anti-interference of the system are not guaranteed[14-15].

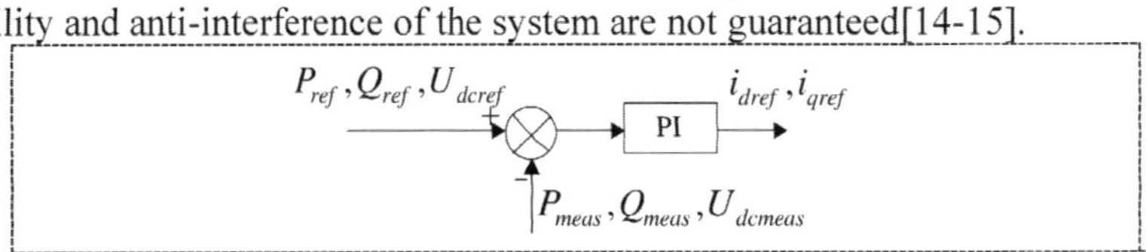

Figure 1. The traditional outer loop controller

In this paper, the improved system level control is adopted in view of the above problems, and the concrete structure is shown in Fig. 2. The improvement 1 is aimed at the improvement of the original outer ring controller, and the improvement 2 is an additional control for the design of abnormal conditions such as the transient fault, like the large disturbance.

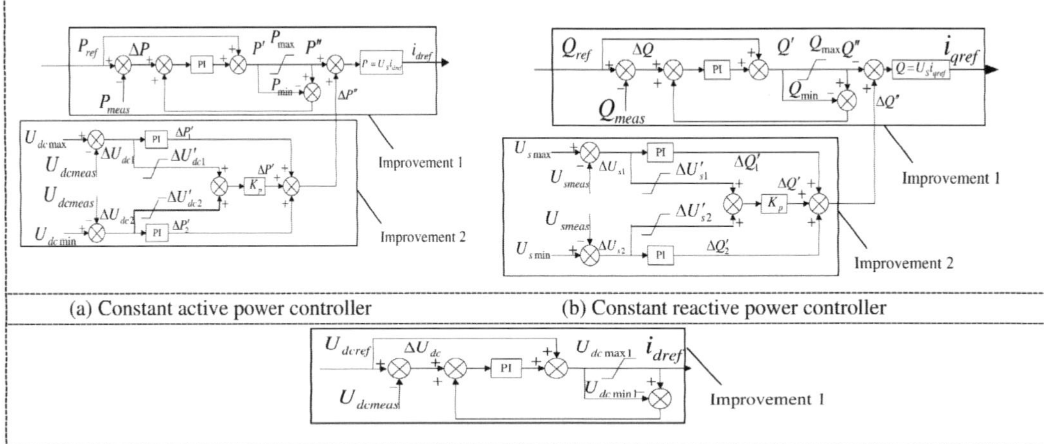

Figure 2. The improved external loop controller

The improved constant active power controller increases the closed loop control compared with the traditional outer loop constant active power controller, and adds transient fault control, as shown in Fig. 2 (a) improvement 1 and improvement 2. Among them, $U_{dc\max}$, $U_{dc\min}$ is the allowable upper and lower limits of DC voltage, the general allowable voltage fluctuation is less than 5% of the rated value . The DC upper limit is 1.05p.u, the lower limit is 0.95p.u, and the upper and lower limits of active power limiting link are 1p.u and -1p.u respectively.

First, the control principle of the improvement 1 is analyzed in detail. Compared to P_{ref} and P_{meas}, the unbalance value ΔP can be obtained, through the PI link the ΔP can be changed into P', through the limiting link, the P' can be changed into P''. Under the steady state, $P'=P''$, when the system is disturbed, P' may be higher or lower than the limit, when the P' is beyond the limit, the system outputs the limit value directly, and the system is normally output when the P' is not exceeded. At this point, $P' \neq P''$, through the adder, $P'-P''$ will be fed back to the input end, and compared with ΔP , the difference will be used to eliminate the disturbance through the PI controller, thus i_{dref} can maintain stable output.

Secondly, the control principle of the improvement 2 is analyzed in detail. Compared to the $U_{dc\max}$ and U_{dcmeas}, ΔU_{dc1} can be obtained, compared $U_{dc\min}$ with U_{dcmeas}, ΔU_{dc2} can be obtained. On the one hand, ΔU_{dc1} and ΔU_{dc2} can be changed into $\Delta P_1'$ and $\Delta P_2'$ through the PI controller, on the other hand, ΔU_{dc1} can be changed into $\Delta U_{dc1}'$ through the limiting link(-1~0), ΔU_{dc2} can be changed into $\Delta U_{dc2}'$ through the limiting link(0~1). $\Delta U_{dc1}'$ is summed with $\Delta U_{dc2}'$, and $\Delta P'$ can be obtain through Proportional link. At last, $\Delta P''$ can be get through the sum of $\Delta P_1', \Delta P_2'$ and $\Delta P'$, and $\Delta P''$ will be send to improvement 1 in order to adjust the i_{dref} value.

Take the case of the transient fault which occurs at the AC side of the inverter, due to lower active power transmission from receiving end, and unchanged active power transmission from sending end, the surplus active power makes the DC voltage increase. In order to solve this problem, the improvement 2 is created to control the DC voltage. Once the DC voltage is too high, the PI controller will output negative active power, then sum with active power from improvement 1, thereby the receiving end active power will be reductive, eventually the increase of DC voltage is suppressed.

In addition, improvement 2 can be used to suppress DC voltage in transient fault, while the traditional controller can convert constant active power and constant DC voltage converter to control the transient fault effect. In comparison, the additional improvement 2 belongs to direct control, which control more quickly and adjust more conveniently. At the same time, due to the introduction of additional control, the complexity of the control is greatly increased compared with the original outer loop controller's. However, the additional control has the advantages of simple structure, which can decrease the system complexity compared to the complexity of the mutual conversion control of the constant active power and the constant current voltage converter in the control of transient faults. In addition, the range of DC voltage fluctuation controlled by additional control is less than the range of traditional transient control. Therefore, when transient fault is very serious, it is necessary to transform the constant active power and constant DC to replace the improvement 2.

The improved constant reactive power controller is similar to the improved constant active power controller. It also increases the closed loop control, and adds transient fault control, as shown in Fig. 2(b) improvement 1 and improvement 2. Among them, $U_{s\max}$, $U_{s\min}$ are the upper and lower limits of AC voltage respectively which are the same as those of DC voltage limits, and $Q_{\max}$, $Q_{\min}$ are the upper and lower limits of reactive power limit which are 0.5p.u and -0.5p.u respectively.

Compared with Fig. 2(a) and Fig. 2(b), the improved constant reactive power controller is basically the same as that of the constant active power controller. It shows that the specific control principle is consistent with the constant active power controller. Only the control volume is converted from active power and DC voltage to reactive power and alternating voltage, and the function is also converted to suppress the fluctuation of AC voltage. Therefore, this article will not repeat it.

The constant DC voltage controller used in this paper is shown in Fig. 2(c), in which the upper and lower limits $U_{dc\max 1}$, $U_{dc\min 1}$ of the DC voltage limiting link are 1.1p.u and -1.1p.u respectively. Because it is aimed at the control of DC voltage, there is no improvement 2, and the function of the improvement 1 is similar to the functions of the above two controllers. It is mainly used to suppress the DC voltage fluctuation under abnormal conditions.

3. Simulation and analysis of operation characteristics of MMC-HVDC system

3.1. A comparative study of steady state stage

In order to verify the superiority of the improved control strategy adopted in this paper, two sets of MMC-HVDC systems are designed, one of which adopts the original control strategy and the other one adopts the improved control strategy. The steady-state performance of the above MMC-HVDC system is simulated to study their operation characteristics under different working conditions and small disturbance states. The operation states of both of them are compared in different states, and the stability and anti-interference of the improved control strategy are analyzed. The details of the control in the steady state are shown in Table 1.

Table 1. Steady state stage simulation control details

Time/s	Action details
1.5-1.64s	The AC voltage of the rectifier is reduced by 0.1 p.u.
2.0s	Active power reference power is reduced by 0.1 p.u.
2.5s	Reactive power of reactive power is reduced by 0.1 p.u.
3s	The inverter side DC reference voltage decreases by 0.05 p.u.

In accordance with the control details given in the above table, the steady-state simulation of the MMC-HVDC system is carried out. The simulation results are shown in Fig. 3.
Sections should be numbered with a dot following the number and then separated by a single space:

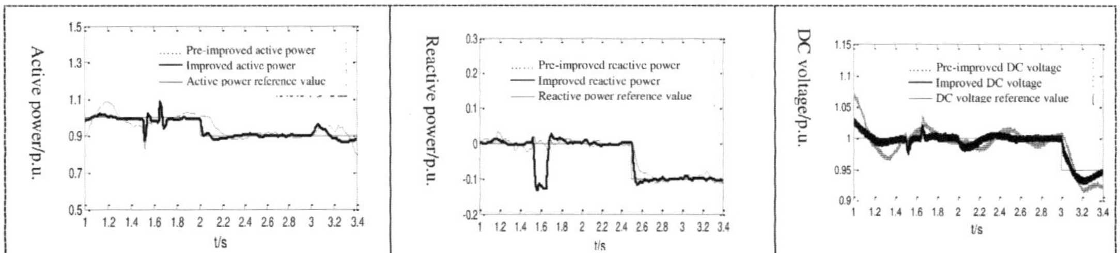

(a) Rectifier side active power (b) Rectifier side reactive power (c) Inverter side DC voltage

Figure 3. Steady state stage waveform

According to Fig. 3, the changes of the system before and after the improvement are analyzed.
(1) Active power: as shown in Fig. 3 (a), in 1.5s-1.64s, the sending side AC voltage drops, and the active power oscillates about 0.18p.u. before the improvement, oscillation time is about 0.3s, after the improvement, the active power oscillates about 0.1p.u., oscillation time is about 0.2s. At 2s, the operation condition changes, the reference value of active power is reduced by 0.1p.u. Before the improvement, the active power real value is hard to follow reference value until 2.2s. The actual value of the improved active power changes with the reference value in time, and the waveform of them is almost coincided. At 2.5s, the reference value of reactive power is reduced by 0.1p.u. No matter before or after improvement, the active power is not changed. At 3s, the DC voltage is reduced by 0.05p.u, the active power changes considerably before the improvement, the recovery time is longer, and the

improved system is stable after 0.2s. The above results show that the improved control strategy has greatly improved the control of the active power, the amplitude of the system oscillation and the recovery time.

(2) Reactive power: as shown in Fig. 3 (b), when the AC voltage is decreasing from 1.5s to 1.64s, before the improvement, the oscillation is very frequent only on the basis of its own regulation mechanism. After adopting the improved control strategy, the reactive power decreases by 0.1p.u under the action of the improved control strategy, thus ensuring the stability of active power and DC voltage. The change of the active power reference value at 2s has no effect on the reactive power. At 2.5s, the reference value of reactive power is reduced by 0.1p.u. Before the improvement, the real value of reactive power tracks the reference value slowly, using 0.3s to reaches stability. After improvement, the actual value of reactive power follows the reference value in time, and almost coincides. At the same time, the change of DC voltage has no effect on reactive power. The above comparison shows that the improved control strategy has a significant effect on reactive power control.

(3) DC voltage: as shown in Fig. 3 (c), when the AC voltage is reduced, the fluctuation of DC voltage is larger before improvement, and the fluctuation amplitude and time are improved after using the improvement 1 and 2. When the reference value of active power is reduced, the DC voltage is reduced, however, the DC voltage amplitude of the fluctuation and the recovery time has improved after improvement. The decrease of reactive power has no effect on the DC voltage. When the DC voltage is reduced, the actual value of DC voltage changes slower and the stability time is more than 0.4s before improvement. After improvement, the actual value can track the reference value in time. It shows that the control effect of the DC voltage is obviously adopted after the improved control strategy.

Through the comparison of simulation results before and after the improvement of control strategy, a conclusion can be get: the improved system can not only make the decoupling control of active power and reactive power, but also can reduce the oscillation amplitude, shorten the oscillation time, make the system reliable, enhance the system stability and anti-jamming.

3.2. A comparative study of transient stage

The three-phase short circuit module is set at the AC side of the inverter, which is put in 3.5s and excised at 3.62s. The specific simulation results are shown in Fig. 4.

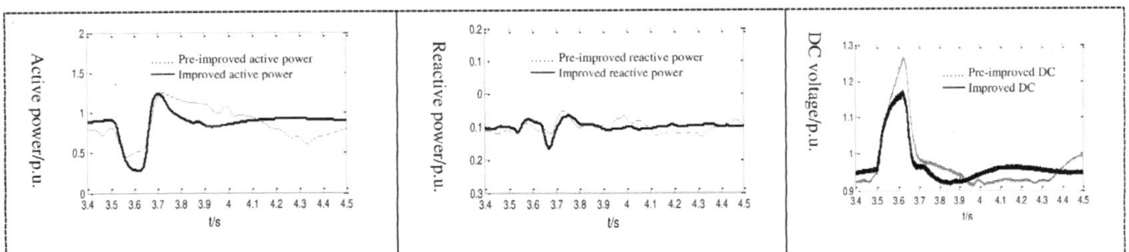

(b)Rectifier side active power (b) Rectifier side reactive power (c) Inverter side DC voltage

Figure 4. Transient state stage waveform

After analyzing the Fig. 4(a), a result can be obtained, when the inverter side occurs three-phase fault, the rectifier side transmission active power is 0.5p.u, the inverter side AC voltage drop to zero, and the active demand drop to 0, at this time, the rectifier side transmission active power is larger than the inverter side active power, the demand imbalance on both sides of the active power will lead to the increase of DC voltage finally, causing damage to the power electronic devices. After using the improved control strategy, the active power of the rectifier side is reduced to 0.25p.u under the action of the controller, which also reduces the harm of the active power imbalance. After the fault is removed, the active power before the improvement always fluctuates at the stable value, but the improved active power is recovered to the stable state after about 0.3s.

After analyzing the Fig. 4(c), a result can be obtained, when three-phase faults occurs, the DC voltage controller works, and controls the maximum amplitude of DC voltage to be 1.16p.u, reduces 0.1p.u.

than that of pre-improved DC voltage amplitude, so after the fault, the improved DC voltage recovery effect is obviously better than the original.

By comparing the simulation results, we can see that the improved control strategy adopted in this paper can reduce the transmission of the active power of rectifier side in time, and control the DC voltage within the allowable range, so as to ensure the safety of power electronic devices. After the fault is removed, the improved system will quickly return to the steady state and there is no obvious fluctuation. The degree of fluctuation and the recovery time of the fault are far less than the time before the improvement, and can be restored to the steady state in time.

4. Conclusion

In this paper, an improved control strategy is adopted in the MMC-HVDC system, which solves the problem of low anti-interference and unstable control in the past. On the platform of MATLAB/Simulink, a simulation model of two side MMC-HVDC system is built. The steady state and operation characteristics of transient stage before and after adopting the improved control strategy are compared.

According to steady state, in the stationary case, the outer loop control is change into closed loop from open loop, which is a certain degree of delay in the control response speed, but the control effect is remarkable. To a certain extent, the improved system can enhance the ability of interference suppression. In the abnormal condition of the system, it can reduce the amplitude of oscillation, shorten the recovery time of system after disturbance, and has good stability and anti-interference.

(1) For transient conditions, the improved control strategy can start without starting conversion control mode, reduce the transmission of rectifier active power in time, so as to effectively suppress the DC voltage increased, and also has a good protective effect on power electronic devices. In addition, the fault recovery time has also been significantly improved.

(2) After using the improved control strategy, although the complexity of MMC-HVDC system control is increased a little, the reaction speed control strategy is reduced, but the advantages are great, the improved system can be more efficient to control the DC system, and provide a more effective basis platform in the research on operational characteristics of AC/DC system for the following research.

5. References

[1] TANG Guangfu, HE Zhiyuan, PANG Hui. "Research, application and development of VSC-HVDC engineering technology", *Automation of Electric Power Systems*, vol.PS-37, no.15, pp.3-14, Feb. 2013.

[2] LI Gengyin, LV Pengfei, LI Guangkai, et al. "Development and prospects for HVDC light", *Automation of Electric Power*, vol.PS-27, no.4, pp.77-81, Mar. 2003.

[3] Xu Feng, Xu Zheng. "Hybrid HVDC system based on LCC and FHMMC", *High Voltage Engineering*, vol.PS-40, no.8, pp.2520-2530, May. 2014.

[4] MA Chenglian, PAN Wenming, YAO Tianliang, et al. "Research on control strategy of VSC-HVDC in AC power grid", *Journal of Northeast Dianli University*, vol.PS-35, no.6, pp.26-32, Feb. 2015.

[5] ZHU Ling, FU Xiaowei, HU Xiaobo, et al. "Model predictive control of modular multilevel converter for HVDC system", *Power System Protection and Control*, vol.PS-42, no.8, pp.1-8, Apr. 2014.

[6] TANG Guangfu, HE Zhiyuan, PANG Hui. "Research, application and development of VSC-HVDC engineering technology", *Automation of Electric Power Systems*, vol.PS-37, no.15, pp.3-14, Feb. 2013.

[7] YANG Xiaofeng, ZHENG Trillion Q. "A novel universal circulating current suppressing strategy based on the MMC circulating current model", *Proceedings of the CSEE*, vol.PS-32, no.18, pp.59-65, Feb. 2013.

[8] FELDMAN R, TOMASINI M, AMANKWAH E, et al. "A hybrid modular multilevel voltage source converter for HVDC power transmission", *IEEE Transactions on Industry Applications*, vol.PS-49, no.4, pp.1577-1588, Jan. 2013.

[9] JIANG Guanqian, LI Zhiyong, YANG Huixia, et al. "Research review on topological structure of flexible HVDC system", *Power System Protection and Control*, vol.PS-43, no.15, pp.145-153, May. 2015.
[10] YANG Xiaofeng, ZHENG Qionglin, XUE Yao, et al. "Review on topology and industry applications of modular multilevel converter", *Power System Technology*, vol.PS-40, no.11, pp.1-10, Jun. 2016.
[11] XING Changda, GUO Jiahu, ZHU Chengjie, et al. "Design of the sub-module controller of the MMC control system", *Power System Protection and Control*, vol.PS-44, no.3, pp.114-121, Feb. 2016.
[12] TANG Geng, XU Zheng, XUE Yinglin. "A LCC-MMC hybrid HVDC transmission system", *Transactions of China Electrotechnical Society*, vol.PS-28, no.10, pp.301-310, Jan. 2013.
[13] YAO Zhiqing, ZHANG Qun, CHEN Peng, et al. "Research on fault diagnosis for MMC-HVDC systems", *Protection and Control of Modern Power Systems*, vol.PS-1, no.1, pp.79-85, Mar. 2016.
[14] XU Zheng, TU Qingrui, GUAN Minyuan, et al. "Voltage source converter based HVDC power transmission systems", *Beijing, China: China Machine Press*, 2013, p.17-19.
[15] Tang Guangfu. "VSC-HVDC technology", *Peking: China Electric Power Press*, 2010, p.30-36.

Acknowledgement
This work was supported by National Key Research and Development Program of China (No. 2017YFB0902000) and National Natural Science Foundation of China (No.U1766204）

IOP Conf. Series: Journal of Physics: Conf. Series **1072** (2018) 012016 doi:10.1088/1742-6596/1072/1/012016

Performance Research on a Twin Screw Expander Applied to a Power Generation System for Recovering Waste Pressure Energy of Natural Gas

S Yao[1], Y Zhang[2] and H Yang[3]

[1] School of Architecture, Tianjin University, Tianjin 300072, PR China

[2] School of Environmental Science and Engineering, Tianjin University, Tianjin 300072, PR China

[3] School of Economics and Management, Tianjin University of Science and Technology, Tianjin 300457, PR China

yaosheng@tju.edu.cn

Abstract. The twin screw expander, applied to a power generation system for recovering waste pressure energy of natural gas (NG), is reliable. It has lower capital costs and maintenance performance compared to the others. In this paper, a computational procedure based on the Benedict-Webb-Rubin (BWR) equation is established. Meanwhile, the experiment is carried out to study the performance of a twin screw expander prototype. The results shows that the twin screw expander has an optimum rotational speed where the specific shaft work, efficiency reach their peaks when delivered a certain gas, and the optimum rotational speed reduces with the decrease of the inlet pressure.

1. Introduction

Natural gas (NG) is mainly transported over long distances through pipeline network systems (corresponding to 67.5% of the global gas trade [1]). It is necessary to transport NG at high pressures (about 10-12MPa [2]) to cut the cost and energy demand. However, the pressure of NG must be significantly reduced to lower levels at pressure reduction stations (PRSs) before it is supplied to utilization sites [3]. Currently, most PRSs employ pressure reducing valves (PRVs) to accomplish pressure reduction and control outlet pressure [4]. It dissipates a considerable amount of available exergy of pressurized gas which can be harnessed to generate work in the expansion process. Many researchers have begun to paid attention to the energy recovery from a NG PRS.

It is possible to convert the available decrease of physical exergy into mechanical work by means of an expander substituting for a PRV [5]. The generated work may be converted to electricity or provided for industrial direct applications.

Although numerous theoretical analyses and evaluation methodologies for various NG expansion systems are widely developed, there are little industrial application and experimental research. In this paper, a power generation system for recovering waste pressure energy of NG is presented and a computational procedure based on the Benedict-Webb-Rubin (BWR) equation is established. Meanwhile, the experiment is carried out to study the performance of a twin screw expander prototype.

2. System description

The experiment used compressed air to substitute for NG as working fluid considering that NG has inflammable, explosive, toxic and other dangerous characteristics. The schematic diagram of the twin screw expander performance test is shown in figure 1. It primarily includes three parts: inlet and exhaust system, oil lubrication system, data measurement and acquisition system.

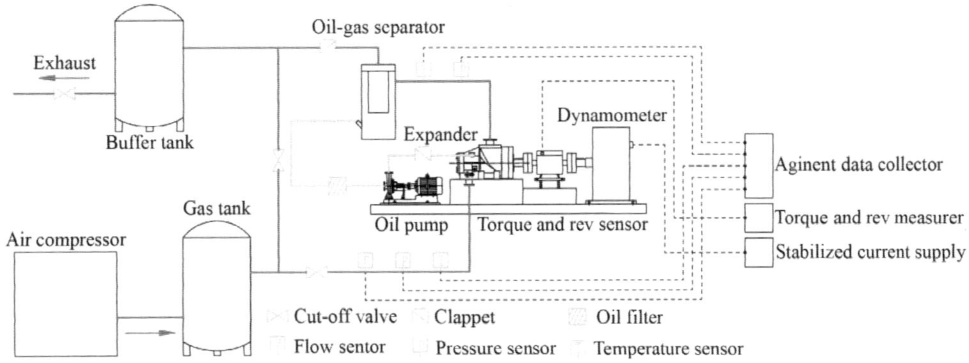

Figure 1. Schematic diagram of the twin screw expander performance test.

3. Computational procedure

The BWR equation of state, which has high accuracy in the application of hydrocarbons gases, was selected for calculating thermodynamic parameters of NG. The fundamental structure of the BWR equation for real gas is as follows:

$$p=\frac{RT}{v}+\left(B_0RT-A_0-\frac{C_0}{T^2}\right)\frac{1}{v^2}+(bRT-a)\frac{1}{v^3}+\frac{a\alpha}{v^6}+\frac{c\left(1+\gamma/v^2\right)}{T^2}\frac{1}{v^3}\exp\left(-\frac{\gamma}{v^2}\right) \tag{1}$$

where A_0 , B_0 , C_0 , a , b , c , α , γ are all constants related to the kind of matter [6,7]. For the mixture, the mixing rules of constants in the BWR equation are defined as follows:

$$A_0=\left(\sum_i x_i A_{0i}^{1/2}\right)^2 \tag{2}$$

$$B_0=\sum_i x_i B_{0i} \tag{3}$$

$$C_0=\left(\sum_i x_i C_{0i}^{1/2}\right)^2 \tag{4}$$

$$a=\left(\sum_i x_i a_i^{1/3}\right)^3 \tag{5}$$

$$b=\left(\sum_i x_i b_i^{1/3}\right)^3 \tag{6}$$

$$c=\left(\sum_i x_i c_i^{1/3}\right)^3 \tag{7}$$

$$\alpha=\left(\sum_i x_i \alpha_i^{1/3}\right)^3 \tag{8}$$

$$\gamma=\left(\sum_i x_i \gamma_i^{1/2}\right)^2 \tag{9}$$

where x_i is the mole fraction of component I in the mixture.

It was assumed that the whole expansion process was gas single-phase flow. A computational procedure was established to realize the thermodynamic calculation. The flow chart of the computational procedure is illustrated in figure 2.

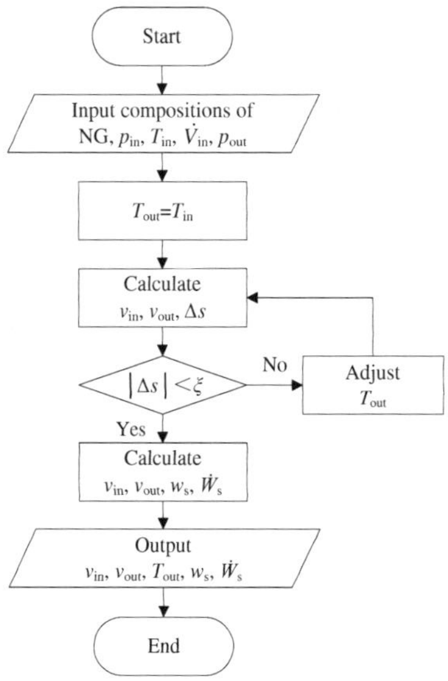

Figure 2. Flow chart of the computational procedure.

4. Results and discussions

The twin screw expander prototype, whose male rotor has 5 teeth and female rotor has 6, is based on the "W" rotor profile. The diameter of the male rotor and female rotor were optimized to 106.8 mm and 84.2 mm, respectively. The axial distance between rotors is 75 mm. The length to diameter ratio of the male rotor is 1.6. A series of experiments were carried out to investigate the performance of the twin screw expander prototype. Throughout the trial process, the expander ran smoothly with low noise and little vibration. Some main performances are analyzed as follows.

4.1. experimental conditions

Taking a set of experimental data as the example, the performance of the twin screw expander at different rotational speeds was analyzed. The air compressor outlet pressure was set between 0.55 MPa and 0.65 MPa, which had contributed to a satisfactory experimental condition that the expander inlet pressure was basically stable at 0.57 MPa and the inlet temperature was approximately 42℃. Meanwhile, the inlet and outlet cut-off valve of the experimental system were fully opened, which implied that the expander back pressure was equivalent to barometric pressure, and the bypass was in a closed state. Moreover, the condition that expander inlet pressure was 0.24 MPa and inlet temperature was 36℃ was used for comparison. The specific values of pressure, mentioned in this paper, refer to gauge pressure.

4.2. Shaft power

Figure 3(a) shows the variation of the shaft power and specific shaft work with the rotational speed. It can be seen that the shaft power increases, whose change rate becomes gradually slower, and the specific shaft work changes with approximate parabola law as the rotational speed increases. The specific shaft work reaches the maximum value of 52.38 kJ/kg at about 2500 r/min. However, the shaft power doesn't reach the maximum value at the rotational speed where has a maximum specific shaft work, and continues to increase. That's because the value of shaft power depends on two aspects: specific shaft work and mass flow rate of working gas.

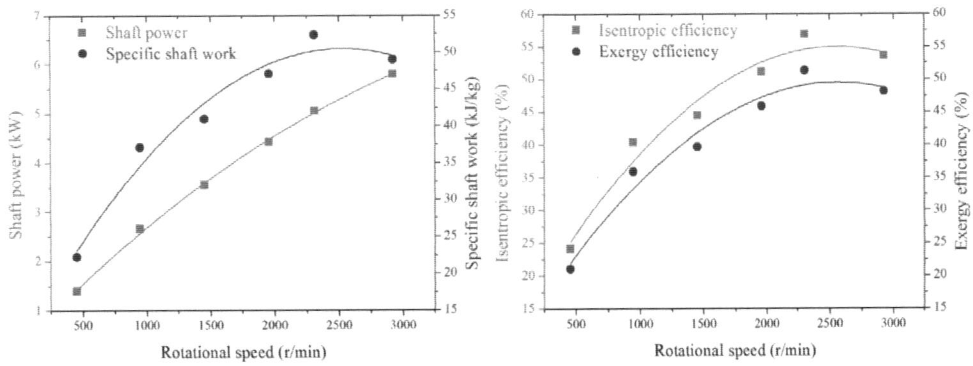

(a) (b)

Figure 3. Effect of the rotational speed on the performance of the twin screw expander.

4.3. Efficiency

Figure 3(b) shows the variation of the isentropic efficiency and exercy efficiency with the rotational speed. It can be seen that both the isentropic efficiency and exercy efficiency change with approximate parabola law as the rotational speed increases, and respectively reach the maximum values of 57% and 51% at about 2500 r/min.

4.4. Inlet pressure

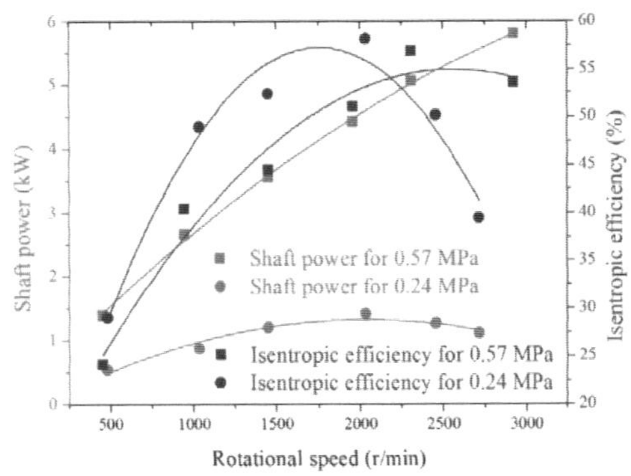

Figure 4. Effect of the inlet pressure on the performance of the twin screw expander.

Figure 4 shows the effect of the inlet pressure on the performance of the twin screw expander. It can be seen that, in contrast with the condition of 0.57 MPa, the shaft power for 0.24 MPa indicates a rough parabolic changes with the increase of rotational speed, and reaches the maximum value at about 2000 r/min. The isentropic efficiency for 0.24 MPa also indicates a parabolic changes as the rotational speed increases, and the difference is its optimum rotational speed is 1750 r/min.

4.5. Shaft power for NG

Under the first test condition and with the same isentropic efficiency, the shaft power for NG was calculated by the proposed computational procedure. The components of NG used are shown in table 1.

Table 1. Compositions of NG.

Compositions	CH_4	C_2H_6	C_3H_8	CO_2	N_2
Mole fraction (%)	88.48	6.68	0.35	3.52	0.97

Figure 5 shows the differences between the shaft power for NG and that for air, the specific shaft work for NG and air also are compared in like manner.

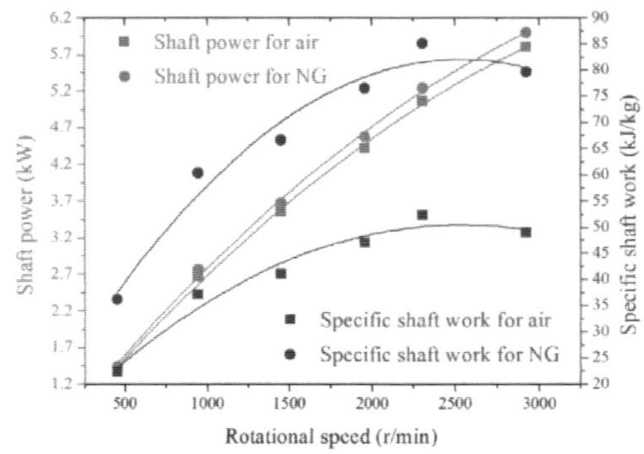

Figure 5. Differences between the shaft power and specific shaft work for NG and that for air.

It can be seen that the shaft power for NG is slightly bigger than that for air and the difference between them are becoming larger as the rotational speed increases. Therefore compressed air can substitute for NG as working fluid to study the performance of the twin screw expander prototype. The specific shaft work for NG is higher than that for air and the difference between them changes with an approximate downward parabola as the rotational speed increases.

5. Conclusions

The main conclusions drawn from the present research are summarized as follows:

(1) The specific shaft work, efficiency change with approximate parabola law as the rotational speed increases. For a certain twin screw expander, there is an optimum rotational speed where the specific shaft work, efficiency reach their peaks when delivers a certain gas. However, the shaft power doesn't necessarily reach the maximum value at that optimum rotational speed.

(2) The optimum rotational speed of the twin screw expander, where the shaft power reached its peak, is not less than the rotational speed where the efficiency is at its maximum.

(3) For the twin screw expander with a certain structure volume ratio, the optimum rotational speed where the isentropic efficiency reaches its peak reduces with the decrease of the inlet pressure.

(4) In reality, compressed air can substitute for NG as working fluid to study the performance of the twin screw expander prototype, which can be applied to a power generation system for recovering waste pressure energy of NG.

6. References

[1] Petroleum British (March 2017) BP statistical review of world energy [Online] Available: https:// www.bp.com/en/global/corporate/energy-economics/statistical-review-of-world-energy.html

[2] Tan H Zhao Q Sun N and Li Y 2016 Proposal and design of a natural gas liquefaction process recovering the energy obtained from the pressure reducing stations of high-pressure pipelines *Cryogenics* (2016)80 pp 82-90

[3] Andrei I Valentinb T Cristinac T and Niculaed T 2014 Recovery of wasted mechanical energy from the reduction of natural gas pressure *Procedia Eng.* 1(2014)69 pp 986-990

[4] Unar I N Aftab A and Abro A M 2015 Estimation of Power Production Potential from Natural Gas Pressure Reduction Stations in Pakistan Using ASPEN HYSYS *Mehran Univ. Res. J. Eng. Technol.* 3(2015)34 pp 1-8

[5] Kostowski W J and Usón S 2013 Thermoeconomic assessment of a natural gas expansion system integrated with a co-generation unit *Appl. Energ.* 1(2013)101 pp 58-66

[6] Su C, Tan L and Liu G 1987 *Advanced engineering thermodynamics* (Beijing: Higher Education Press)

[7] Tong J 2008 *Fluid thermophysical properties - basic theory and calculation* (Beijing: Sinopec press)

Acknowledgments

This study obtained great supports from the financial supports by the 973 National Key Basic Research Program of China (No. 2015CB251403).

IOP Conf. Series: Journal of Physics: Conf. Series **1072** (2018) 012017 doi:10.1088/1742-6596/1072/1/012017

Genetic Algorithm Based Economic Optimization of Microgrid Including Multi-Energy Supply System

Shujian Zhao[1], Chenglian Ma[1], Li Sun[1]*, Yuanzhu Meng[1], Wei Wu[2] and Luheng Huang[1]

[1] School of Electrical Engineering, Northeast Electric Power University, Jilin, Jilin 132012, China

[2] State Grid Fushun Electric Power Supply Company, Fushun, Liaoning, 113005, China

* Li Sun: sunlinedu@163.com

Abstract. It is necessary to reduce CO_2 emission and improve energy efficiency saving in microgrid power generation scheduling. In this context, both cost and benefit are analyzed. Considering the CO_2 emission, the energy efficiency saving and some constraint conditions, an economic model including the power supply and fuel gas supply is provided. The model is solved by genetic algorithm. The optimization results of calculation example show that the model provided is feasible and reliable. And it can also be applied to a large-scale power grid.

1. Introduction

With the increasing awareness of energy saving and environmental protection, distributed generation (DG) is paid more and more attention. Microgrid combines all kinds of distributed power, load, energy storage unit and control device to form a single controllable unit, providing electricity and heat energy to users at the same time, realizing cogeneration. Microgrid has become an effective technique for managing DG in smart grid, and it has been widely studied for the operation and control of microgrid. Therefore, it is necessary to study the microgrid [1], which is an effective supplement and strong support for the large power grid. Microgrid is clean and efficient. It can improve energy utilization and reduce environmental pollution [2].

To improve the energy efficiency and reduce CO_2 emission, the two aspects of cost and benefit are in consideration in this paper. Considering the emissions, an economic model including multi-energy supply is provided. The model is solved by genetic algorithm [3-4].

The analysis results show that the incorporation of microgrid into power grid can improve energy efficiency, reduce costs, and reduce CO_2 emission. It can also reflect the effectiveness and feasibility of economic operation of microgrid [5].

2. Economic Analysis of Microgrid

The economic analysis mainly includes benefit analysis and cost analysis. The difference value between benefit and cost can be used as an instruction of economic analysis.

The benefit of the microgrid mainly includes the benefit of the distribution network, the benefit of the user, the benefit of the social and so on. The benefit of distribution network refers that microgrid can improve power supply reliability, support voltage sags in distribution network, and delay or avoid equipment investment in distribution network. The benefit of the user refers that microgrid can reduce the energy cost of the thermal and thermal natural gas, improve the reliability of the power supply and provide the auxiliary service. The benefit of the social refers that microgrid can improve the utilization of energy and reduce the CO_2 emission.

The cost of microgrid mainly includes initial investment cost, the cost of operation and maintenance, the cost of fuel, the cost of emission penalty and so on. The initial investment cost is the microgrid construction cost. The cost of operation and maintenance includes network loss, cost of start and stop, cost of electricity purchase and so on. The cost of fuel mainly includes the cost of natural gas and fuel. The cost of emission penalty mainly includes the penalty cost of CO_2 emission released by power generation.

3. Economic Optimization Model of Microgrid

3.1. Microgrid Structure

In microgrid, five types of power supply are put into use, including Photovoltaic (PV), Wind Turbine (WT), Hydropower (HP), Microturbine (MT) and Fuel Cell (FC), as shown in Figure 1. The relationship of supply and demand can be seen clearly.

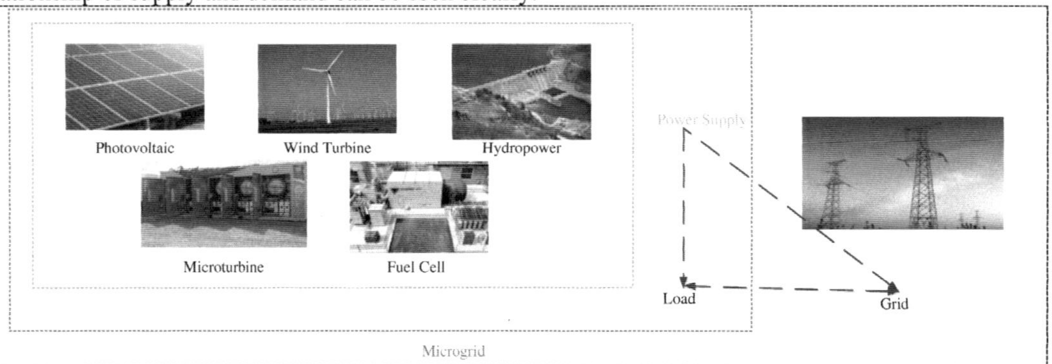

Figure 1. The structure of microgrid.

3.2. Objective Function and Constraint Condition

Then a microgrid model is established with the integration of power supply, thermal supply and gas supply. In order to maximize the benefit of microgrid and minimize the total cost of microgrid, the objective of establishing a microgrid economic model for grid operation is to be built. Considering the complexity of the model and the uncertainty of thermal energy demand, only the cost and benefit of electricity and natural gas supply are in considered, and cost and benefit of thermal supply is not considered in this paper.

The objective function of the economic model established in this paper is

$$\min\left\{\sum_{n=1}^{N} C_n \frac{R(1+R)^{lc_n}}{(1+R)^{lc_n}-1} + C_{BATTRY}\frac{R(1+R)^{lc_{BATTRY}}}{(1+R)^{lc_{BATTRY}}-1} + \sum_{n=1}^{N} C_{mn}P_n + \sum_{n=1}^{N} E_n\lambda \right. \\ \left. +\alpha P_b - \beta P_s + \gamma T + \delta G_{MT} + \delta G_{FC} + \delta G_p - \varepsilon G_s\right\} \tag{1}$$

where, n=1~N, where N is the total number of power supply in the microgrid; C_n is the initial cost for the n power supply; lc_n is the lifecycle of the n unit. R is the rate of interest; $C_{BATTARY}$ is the initial cost of energy storage device; $lc_{BATTARY}$ is the lifecycle of energy storage device; C_{mn} is the maintenance cost of the n unit. P_n is the power produced per hour of the n unit. E_n is CO_2 emission from the n unit. λ_2 is emission penalty price. α is power price purchasing from the grid; P_p is power purchasing from the grid; β is power price selling to the grid; P_s is power selling to the grid; γ is the cost of each start-stop. T is the start-stop frequency of the power supply; δ is gas price purchasing from the microgrid; G_{MT} and G_{FC} are the natural gas required for the produced by MT and FC; G_p is the natural gas purchasing from the grid; ε is gas price selling to the microgrid; G_s is the gas selling to the microgrid.

The equality constraint of the model is the power balance constraint, that is

$$P_d = P_{PV} + P_{WT} + P_{HP} + P_{MT} + P_{FC} + P_p - P_s - P_{loss} \quad (2)$$

where, P_d is the total power demand per hour. P_{PV}, P_{WT}, P_{HP}, P_{MT} and P_{FC} are the power produced by PV, WT, HP, MT and FC, respectively. P_{loss} is the power loss of the microgrid.
The inequality constraints of the model include power upper and lower limit constraint, CO_2 emission constraint and start-stop times constraint. The upper and lower limit constraint is

$$P_n^{\min} \le P_n \le P_n^{\max} \quad (3)$$

where, P_n^{min} and P_n^{max} are the upper and lower limit of the output of the n unit, respectively. The emission constraint of CO_2 is as follows:

$$\sum_{n=1}^{N} E_n \le E_{\max} \quad (4)$$

where, E_{max} is the maximum emission limit of CO_2. The start-stop frequency constraint is as follow:

$$T < T_{\max} \quad (5)$$

where, T_{max} is the maximum frequency limit of the microgrid.

4. Solving Method

Genetic algorithm is one of the evolutional computing techniques developed in recent years. Genetic algorithm is a computational model of biological evolution process of natural selection and genetic mechanism simulating Darwin's theory of biological evolution. It's a way to search for optimal solution by simulating natural evolution process. Genetic algorithm starts from a population representing potential problems, and a population is composed of a number of individuals (individual) encoded by genes. Each individual is actually an entity with the characteristics of a chromosome. Chromosome is the main carrier of genetic material, that is, the collection of multiple genes. Its internal expression is a combination of genes, which determines the external expression of individual shape [6].
The research is based on daily load distribution. According to the power supply of microgrid shown in Figure 1 and the microgrid economic model built above, the formula (1) can be simplified that the objective function contain only cost of power generation and CO_2 emission. The power balance constraint, power upper and lower limit constraint, emission constraint of CO_2 and start-stop frequency constraint are in considered. Genetic algorithm can be achieved by Matlab, and the simulation calculation is carried out. The flowchart of genetic algorithm is given in Figure 2. The load in microgrid is predicted and analyzed. The daily load curve is shown in Figure 3. The daily load curves of summer, winter, spring and autumn are expressed by curves 1, 2, 3 respectively.

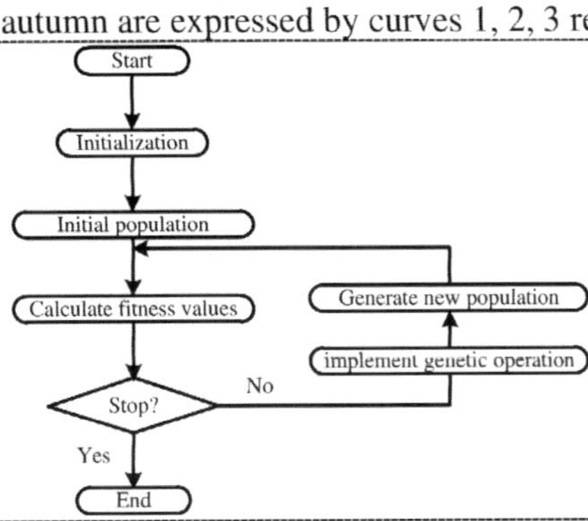

Figure 2. The flowchart of genetic algorithm.

IOP Conf. Series: Journal of Physics: Conf. Series **1072** (2018) 012017 doi:10.1088/1742-6596/1072/1/012017

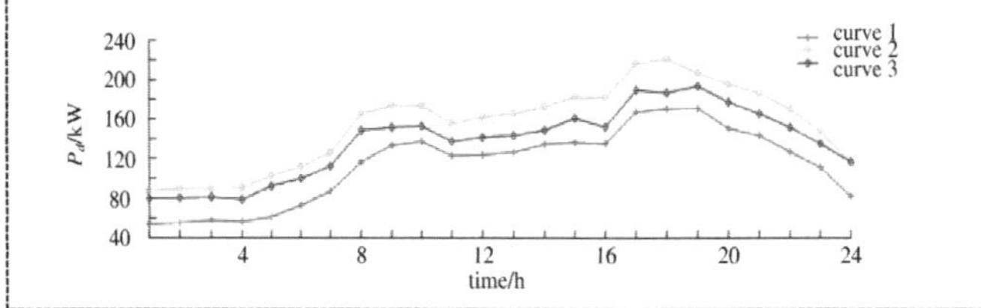

Figure 3. Predicted load curve in different seasons.

5. Case Study

5.1. Parameter Settings

The output power of PV will be greatly influenced by sunshine intensity in different seasons. Therefore, the output power of PV should also be analysed in summer, winter, spring and autumn. As shown in Figure 4, the curves 1, 2, 3 represent the output power of PV in summer, winter, spring and autumn.

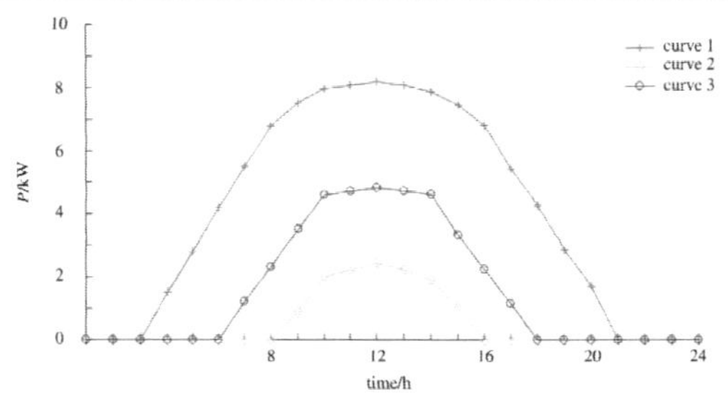

Figure 4. Daily output power of PV.

The output power of WT is influenced by wind. The output power of HP is influenced by water. Therefore, the output power of WT and HP are assumed as a constant to simplify calculation. The generation power cost of PV, WT and HP are low, and during the generation power, there is nearly no CO_2 emission. Then it is assumed that there is no cost or CO_2 emission during the generation power of PV, WT and HP. The cost and CO_2 emission are only from MT and FC. The parameter settings are shown in Table 1.

Table 1. The parameter settings.

Parameter	Value	Parameter	Value
Output power of WT	3.5kW/h	Output power of HP	2kW/h
CO_2 emission penalty price	0.30yuan/kg	CO_2 emission limit	200kg
CO_2 emission parameter of MT	0.72	CO_2 emission parameter of FC	0.43
Fuel consumption coefficient of MT	0.40m^3/kWh	Fuel consumption coefficient of FC	0.25 m^3/kWh
Maintain cost of MT	0.04yuan/kWh	Maintain cost of FC	0.03yuan/kWh
Power upper limit of MT	330	Power upper limit of FC	25
Power lower limit of MT	30	Power lower limit of FC	5

5.2. Simulation Result

In order to minimize the objective function (total cost), the output power curves of MT and FC after optimization are shown in Figure 4. Curves 1, 2, 3, 4 represent the output power of FC in summer, the output power of MT in summer, the output power of MT in winter, and the output power of MT in spring and autumn, respectively. As shown in Figure 5, curves 1, 2, 3 represent the total cost of microgrid in summer, winter, spring and autumn, respectively.

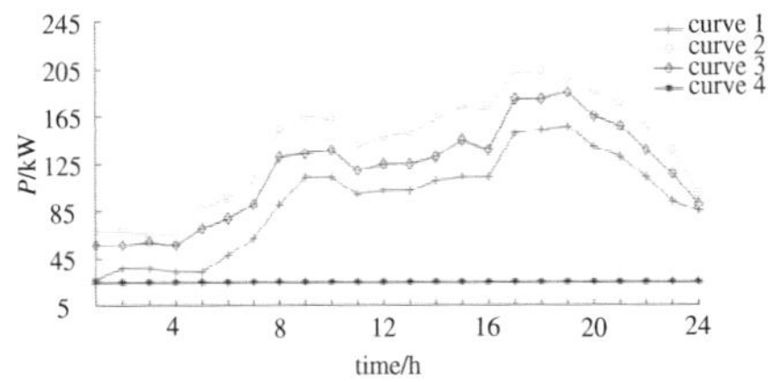

Figure 5. The output power of MT and FC after optimizing.

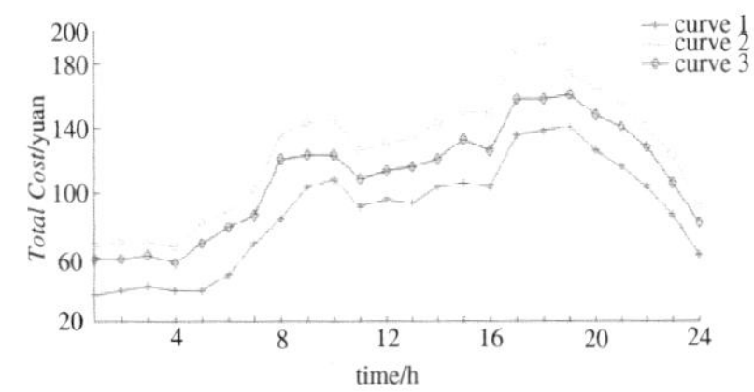

Figure 6. The total cost after optimizing.

6. Conclusion

1) In summer, CO2 emission is reduced by17.8% after optimizing, and then total cost is the lowest. It is because that PV can generate more power than in winter, and generation power of MT is reduced
2) Without considering selling electricity to the grid, the clean energy is firstly to generate electricity, such as PV, WT and HP. And then the lower emission and higher efficiency units are on the turn to generate electricity, such as MT. Finally, the higher emission units are used to generate electricity, such as FC.
3) The model provided in this paper can be used to each multi-energy supply system, including thermal supply. Actually, thermal supply is more popular in winter and further work is currently in progress to forecast the uncertainty of thermal energy demand.

7. References

[1] Zadsar M, Haghifam M R and Larimi S M M 2018 *Approach for self-healing resilient operation of active distribution network with microgrid* (Iet Generation Transmission & Distribution) vol 11 chapter 18 pp 4633-4643
[2] Lin Y, Dong P, Sun X and Liu M 2017 *Two-level game algorithm for multi-microgrid in electricity market* (Iet Renewable Power Generation) vol 11 chapter 14 pp 1733-1740
[3] Vahedipour-Dahraie M, Rashidizadeh-Kermani H, Najafi H R, Anvari-Moghaddam A and Guerrero J M 2017 *Stochastic security and risk-constrained scheduling for an autonomous microgridwith demand response and renewable energy resources* (Iet Renewable Power Generation) vol 11 chapter 14 pp 1812-1821
[4] John B, Ghosh A, Zare F and Rajakaruna S 2017 *Improved control strategy for accurate load power sharing in an autonomous microgrid* (Iet Generation Transmission & Distribution) vol 11 chapter 17 pp 4384-4390
[5] Fichera A, Frasca M and Volpe R 2017 *Complex networks for the integration of distributed energy systems in urban areas* (Applied Energy) **193** 336-345
[6] El-Hendawi M, Gabbar H A, El-Saady G and Ibrahim E N A 2017 *Enhanced MG with optimum operational cost of pumping water distribution systems* (IEEE International Conference on Smart Energy Grid Engineering) pp 137-142

Acknowledgments
This work was supported by National Key Research and Development Program of China (No. 2017YFB0902000) and Research on Global Energy Interconnection technical system.

Heat Transfer Analysis of Blast Furnace Stave Based on Entransy Theory

X Xu[1,2]**, L j Wu**[1]

[1] Department of Thermal Engineering, School of Mechanical Engineering, TongJi University, Shanghai 201804, China
[2] Nantong University Xinglin College, Jiangsu Nantong 226000, China

xinshoux2@126.com

Abstract. Based on entransy theory, the entransy balance equation and the entransy dissipation of blast furnace stave are obtained. The thermal resistance of blast furnace stave is defined. It is proposed to combine the maximum temperature of the hot surface and the thermal resistance to evaluate the heat transfer performance of the cooling stave. They represent the thermal safety and the overall heat transfer performance of the cooling stave respectively. The calculation method of the thermal resistance of blast furnace stave is illustrated with a practical example. The influence of the radius of the water pipe, the water pipe interval and the shape of the water pipe on the maximum temperature and the thermal resistance of the cooling stave is discussed. The results show that the maximum temperature and thermal resistance of the cooling stave have different changing tendency with the influence of different factors. In the optimization design of the blast furnace stave, the design scheme with smaller thermal resistance can be chosen on the premise of ensuring thermal safety, so as to improve the overall heat transfer performance.

1. Introduction

Blast furnace stave is an important cooling equipment installed in the blast furnace. Its heat transfer performance is of great significance to the production safety of the blast furnace and the efficient utilization of energy. Many scholars have made deep research on the heat transfer analysis of the blast furnace stave. The influence of various factors on the temperature distribution of the cooling stave are discussed[1-5]. In most of these studies, the maximum temperature of the cooling stave is used as an index to evaluate the heat transfer performance. However, the maximum temperature of the cooling stave, as a local parameter, can be used to evaluate the thermal safety of the cooling stave, but it does not reflect the overall heat transfer performance of the cooling stave. Therefore, a physical quantity which reflects the overall heat transfer performance of the cooling stave should be sought, together with the maximum temperature of the cooling stave, to be used as the evaluation index of the heat transfer performance of the blast furnace stave.

To reflect the overall heat transfer performance of the system, Guo et al. [6] defined the physical quantity 'entransy' which represents the heat transfer capacity, and proposed the entransy dissipation extremum principle. Many scholars have conducted the research on the optimization of heat transfer with the minimum entransy dissipation as the target [7-13]. In the optimization of a heat conduction process, Cheng et al. [14] proposed the minimum thermal resistance principle. The smaller the equivalent thermal resistance, the better the overall heat transfer performance of the system is. In this paper, the entransy theory is applied to the heat transfer analysis of the blast furnace stave. The

IOP Conf. Series: Journal of Physics: Conf. Series **1072** (2018) 012018 doi:10.1088/1742-6596/1072/1/012018

entransy balance equation and the entransy dissipation of the stave are obtained. On this basis, the thermal resistance of the blast furnace stave is defined. It can be used as the evaluation index of the overall heat transfer performance of the cooling stave. The calculation method of the thermal resistance of blast furnace stave is illustrated with a practical example. The influence of the radius of the water pipe, the water pipe interval and the shape of the water pipe on the maximum temperature and the thermal resistance of the cooling stave is discussed.

2. Entransy theory summary

Entransy is a physical quantity describing heat transfer ability. It is defined

$$E_h = \frac{1}{2} Q_{Vh} T \tag{1}$$

where Q_{Vh} is the thermal energy stored in an object of constant volume, J, and T is the temperature of the object, K.

The dissipation of entransy per unit time and per unit volume, ϕ_h, is called the entransy dissipation function.

$$\phi_h = -\dot{q} \cdot \nabla T \tag{2}$$

where $\dot{q}$ is the heat flux vector, $W \cdot m^{-2}$, and Ñ T is the temperature gradient, $K \cdot m^{-1}$.

For three-dimensional problems, the entransy dissipation of the whole volume is

$$E_{h\phi} = \int_V \phi_h dV \tag{3}$$

On this basis, the equivalent thermal resistance of the three-dimensional heat conduction problem is defined as

$$R_h = \frac{E_{h\phi}}{\dot{Q}_h^2} \tag{4}$$

where $E_{h\phi}$ is the entransy dissipation, $W \cdot K$, and $\dot{Q}_h$ is the total heat flow on the boundary, W.

The minimum thermal resistance principle can be expressed as follows, for the heat conduction problem with certain constraint conditions, if the equivalent thermal resistance of the object is the minimum, the thermal conductivity of the object is the best.

3. The equivalent thermal resistance of the blast furnace stave

The simplified model of the blast furnace stave can be regarded as a three-dimensional steady state heat conduction, which is shown in Figure 1.

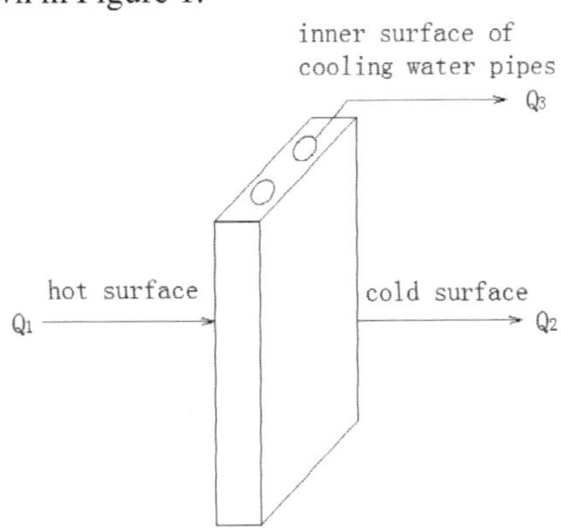

Figure1. Three-dimensional graph of blast furnace stave

The heat flow of each surface is

$$\dot{Q}_i = \int_s \dot{q}_i dA_i \tag{5}$$

where $\dot{q}_i$ is the heat flux on each surface element, $W \cdot m^{-2}$, A_i is the area of each surface, m^2.

The relationship of the heat flow of each surface is

$$\dot{Q}_1 = \dot{Q}_2 + \dot{Q}_3 \tag{6}$$

where $\dot{Q}_1$ is the input heat flow from the hot surface, $\dot{Q}_2$ is the output heat flow from the cold surface, $\dot{Q}_3$ is the output heat flow from the inner surface of the cooling water pipes. The input heat flow from the hot surface is equal to the sum of the output heat flow from the cold surface and the inner surface of the cooling water pipes.

The entransy flux of each surface is

$$\dot{E}_{hi} = \int_s \dot{q}_i T_i dA_i \tag{7}$$

where T_i is the temperature of each surface element, K.

The input entransy flux is not equal to the output entransy flux in the blast furnace stave because of the entransy dissipation in the process of heat transfer. Their relationship is

$$\dot{E}_{h1} = \dot{E}_{h2} + \dot{E}_{h3} + E_{h\phi} \tag{8}$$

where $\dot{E}_{h1}$ is the input entransy flux from the hot surface, W·K, $\dot{E}_{h2}$ is the output entransy flux from the cold surface, W·K, $\dot{E}_{h3}$ is the output entransy flux from the inner surface of the cooling water pipes, W·K, $E_{h\phi}$ is the entransy dissipation of the blast furnace stave, W·K. Equation (8) is the entransy balance equation of three-dimensional steady-state heat conduction of the blast furnace stave. The input entransy flux of the blast furnace stave is equal to the sum of the output entransy fluxes and the dissipated entransy.

The equivalent thermal resistance of the blast furnace stave is equal to the entransy dissipation divided by the square of the heat flow:

$$R_h = \frac{E_{h\phi}}{\dot{Q}_h^2} = \frac{E_{h\phi}}{\dot{Q}_1^2} \tag{9}$$

The thermal resistance of the blast furnace stave can be used as an evaluation index of the whole heat transfer performance. The smaller the thermal resistance, the better the overall heat transfer performance is.

4. Heat transfer analysis of the blast furnace stave

4.1. Calculation of the thermal resistance of a blast furnace stave

The blast furnace of an experiment was selected as the research object. The thickness of the stave is 0.2 m, the width is 0.6 m, and the height is 1 m. The physical properties of cast steel for cooling stave material are as follows: the density is 7800 kg·m^{-3}, the thermal conductivity is 52 W·m^{-1}·K^{-1}, and the specific heat at constant pressure is 486 kJ·kg^{-1}·K^{-1}.

The steady-state heat transfer of blast furnace stave can be regarded as a heat conduction problem. The three-dimensional steady state heat conduction differential equation is

$$\frac{\partial}{\partial x}(\lambda\frac{\partial T}{\partial x}) + \frac{\partial}{\partial y}(\lambda\frac{\partial T}{\partial y}) + \frac{\partial}{\partial z}(\lambda\frac{\partial T}{\partial z}) = 0 \tag{10}$$

where T is the temperature, K, $\lambda(T)$ is the thermal conductivity when the temperature is T, W·m^{-1}·K^{-1}, x, y, z are the directions of the thickness, width and height of the stave respectively.

The heat exchange between the cold surface and the ambient air is the third kind boundary condition. The temperature of the air is 30 °C, and the natural heat transfer coefficient between the cold surface and the ambient air is 11 W·m^{-1}·K^{-1}. The heat exchange between the cooling water and the cooling stave is the third kind boundary condition. The temperature of the cooling water is 30 °C. The calculation result shows that it is the turbulent flow with forced convection heat transfer in the pipe. The convective heat transfer coefficient between the cooling water and the inner surface of the water pipe is[15]

$$h_{wb} = 0.023 v^{0.8} \lambda_w P_r^{0.4} d_1^{-0.2} \upsilon^{-0.8} \tag{11}$$

where d_1 is the equivalent diameter of the cooling water pipe, m, v is the cooling water velocity in the pipe, $m{\cdot}s^{-1}$, λ_w, v and P_r are the thermal conductivity, $W{\cdot}m^{-1}{\cdot}K^{-1}$, Kinematic viscosity, $m^2{\cdot}s^{-1}$, and Prandtl number of the water at 30 °C respectively.

The heat exchange between the gas and the hot surface of the blast furnace stave is complicated. There are two forms of heat exchange, convection and radiation. The composite heat transfer coefficient between the gas and the hot surface of the stave is 232 $W{\cdot}m^{-1}{\cdot}K^{-1}$[16]. The gas temperature in the laboratory furnace is 400 °C.

The three-dimensional heat transfer model of the blast furnace stave was evaluated using the finite element software ANSYS. The temperature field of the cooling stave is obtained by ANSYS calculation.

The entransy flux on the hot surface, cold surface, and inner surface of the cooling water pipes can be calculated as follows:

$$\dot{E}_h = \sum_{i=1}^{n} \dot{q}_i T_i a_i \tag{12}$$

where $\dot{q}_i$ is the heat flux of each node on each surface, $W{\cdot}m^{-2}$, T_i is the temperature of each node on the surface, K, a_i is the area occupied by each node, m^2, and n is the number of nodes of each surface.

The temperature of each node are brought into equation (12) and calculated, the input entransy flux on the hot surface $\dot{E}_{h1}$, the output entransy flux on the cold surface $\dot{E}_{h2}$, and the output entransy flux on the inner surface of the cooling water pipe $\dot{E}_{h3}$ can be obtained respectively. $\dot{E}_{h1}$, $\dot{E}_{h2}$, $\dot{E}_{h3}$ are brought into equation (8), (9) and calculated, the entransy dissipation $E_{h\phi}$ and the thermal resistance of the blast furnace stave R_h can be obtained.

4.2. Effect of the radius of the cooling water pipe on the maximum temperature of the hot surface and the thermal resistance of the stave

The overall structural parameters of the cooling stave are set unchanged. Two cooling water pipes are evenly arranged with the interval of 300mm, and the cooling water velocity is 2m/s. The maximum temperature of the hot surface and the thermal resistance of the stave are calculated with different radius of the cooling water pipe. The results are shown in Figure 2.

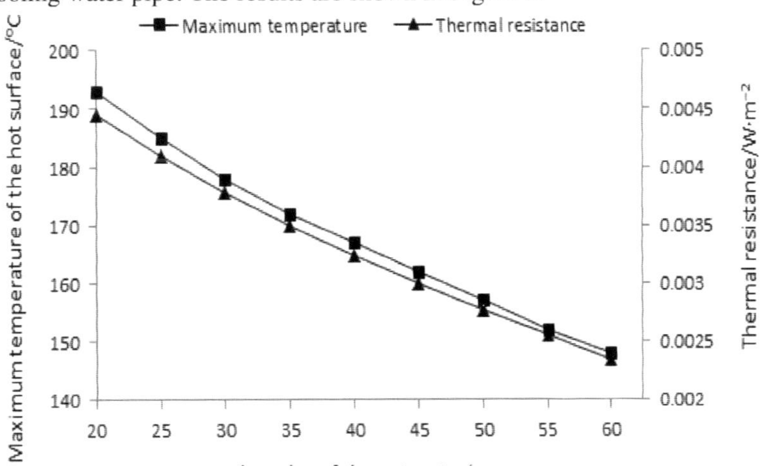

Figure 2. Effect of the radius of the cooling water pipe on the maximum temperature of the hot surface and the thermal resistance of the stave

The maximum temperature of the hot surface and the thermal resistance of the cooling stave decreased rapidly with the increase of the radius of the water pipe. When the radius of the pipe is increased from 20mm to 60mm, the maximum temperature of the hot surface is decreased by 45°C, and the decrease rate is 23.3%. The corresponding thermal resistance decreased by 47.2%. This shows that increasing the radius of the cooling water pipe can effectively improve the thermal safety and the overall heat transfer performance of the cooling stave.

4.3. Effect of the cooling water pipe interval on the maximum temperature of the hot surface and the thermal resistance of the stave

The overall structural parameters of the cooling stave are set unchanged. The cooling water velocity is 2m/s, and the radius of the pipe is 0.02m. The maximum temperature of the hot surface and the thermal resistance of the stave are calculated with different intervals of the cooling water pipe (2 cooling water pipes arranged with the interval of 300mm; 3 cooling water pipes arranged with the interval of 200mm; 4 cooling water pipes arranged with the interval of 150mm; 5 cooling water pipes arranged with the interval of 120mm). The results are shown in Figure 3.

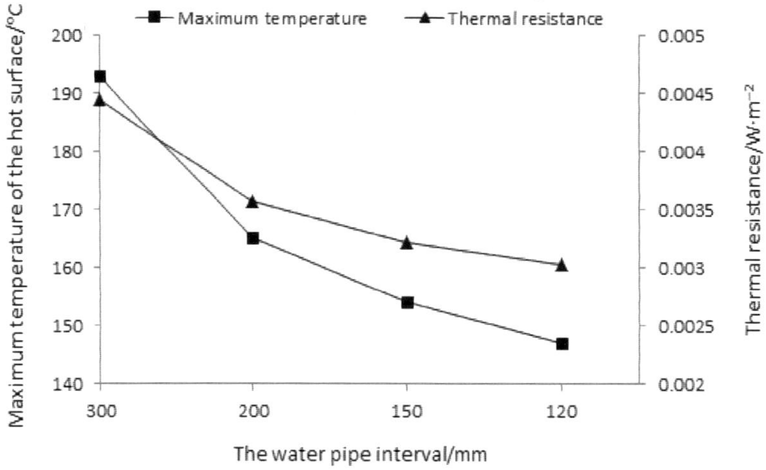

Figure 3. Effect of the cooling water pipe interval on the maximum temperature of the hot surface and the thermal resistance of the stave

The maximum temperature of the hot surface and the thermal resistance of the cooling stave decreased rapidly with the decrease of the water pipe interval. When the water pipe interval is decreased from 300mm to 120mm, the maximum temperature of the hot surface is decreased by 46°C, and the decrease rate is 23.8%. The corresponding thermal resistance is decreased by 32.1%. This shows that decreasing the interval of the cooling water pipe can effectively improve the thermal safety and the overall heat transfer performance of the cooling stave.

In comparison with figure 3, it is found that with the same thermal resistance decline, reducing the cooling water pipe interval can reduce the maximum temperature of the hot surface more effectively, thus improving the thermal safety of the cooling stave; while with the same thermal maximum temperature drop, increase the radius of the cooling water pipe can reducing the thermal resistance more effective, thus improving the overall heat transfer performance of the cooling stave.

4.4. Effect of the shape the cooling water pipe on the maximum temperature of the hot surface and the thermal resistance of the stave

The overall structural parameters of the cooling stave are set unchanged. The 2 cooling water pipes are evenly arranged with the interval of 300mm, the radius of the pipe is 0.02m, and the cooling water velocity is 2m/s. Then the cooling water pipe was changed from circular to elliptical with the same surface area. The maximum temperature of the hot surface and the thermal resistance of the stave are calculated with different ratios of the minor axis to major axis (a/b) of the elliptical pipe. The results are shown in Figure 4.

The maximum temperature of the hot surface and the thermal resistance of the cooling stave increased slowly with the decrease of a/b, and the curve becomes steep when the ratio (a/b) is less than 0.7. When the ratio (a/b) is decreased from 1(circular pipe) to 0.7, the maximum temperature of the hot surface increased by3°C, and the increase rate is 1.6%. The corresponding thermal resistance increased by 4.3%. When the ratio (*a*/*b*) is decreased from 0.7 to 0.5, the maximum temperature of the hot surface increased by 7°C, and the increase rate is 3.6%. The corresponding thermal resistance increased by 5.3%. This shows that changing the shape of the cooling water pipe from circular to elliptical with the same surface area will reduce the thermal safety and the overall heat transfer performance of the cooling stave.

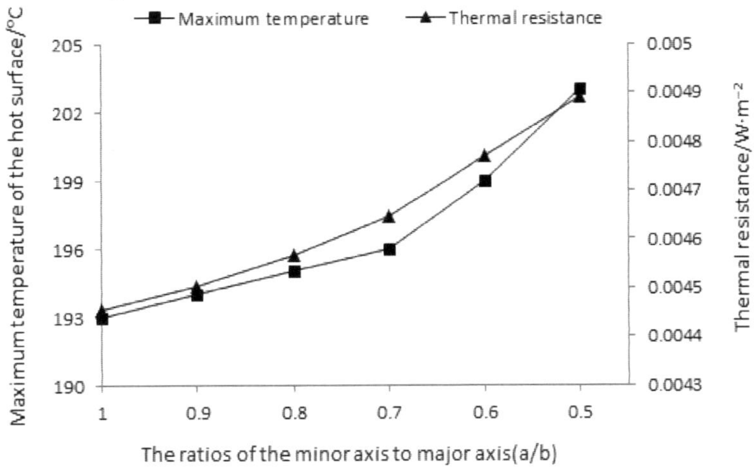

Figure 4. Effect of the shape the cooling water pipe on the maximum temperature of the hot surface and the thermal resistance of the stave

5. Conclusions

In this paper, the entransy theory is applied to the heat transfer analysis of the blast furnace stave. The entransy balance equation and the entransy dissipation of the stave are obtained. The thermal resistance of the blast furnace stave is defined. It can be used as the evaluation index of the overall heat transfer performance of the cooling stave. The smaller the thermal resistance, the better the overall heat transfer performance is. The calculation method of the thermal resistance of the blast furnace stave is illustrated with a practical example. The influence of the radius of the water pipe, the water pipe interval and the shape of the water pipe on the maximum temperature and the thermal resistance of the cooling stave is discussed. The result shows that:

(1) Increasing the radius of the cooling water pipe or decreasing the interval of the cooling water pipe can both effectively improve the thermal safety and the overall heat transfer performance of the cooling stave. The comparative analysis shows that: with the same thermal resistance decline, reducing the cooling water pipe interval can reduce the maximum temperature of the hot surface more effectively, thus improving the thermal safety of the cooling stave; while with the same thermal maximum temperature drop, increase the radius of the cooling water pipe can reducing the thermal resistance more effective, thus improving the whole heat transfer performance of the cooling stave.

(2) Changing the shape of the cooling water pipe from circular to elliptical with the same surface area can increase the maximum temperature and thermal resistance of the cooling stave. It will reduce the thermal safety and the overall heat transfer performance of the cooling stave.

(3) The maximum temperature and thermal resistance of the cooling stave have different changing tendency with the influence of different factors. In the optimization design of the blast furnace stave, the two evaluation indexes should be considered at the same time. The design scheme with smaller thermal resistance can be chosen on the premise of ensuring thermal safety, so as to improve the overall heat transfer performance.

6. References

[1] Qian Z, Cheng H E and Wu L J 2006 Heat transfer analysis of cooling stave based on thermal state experiment *Research on Iron and Steel* **18(5)** 10–13

[2] Wu L J, Zhou W G, Su Y L, et al 2008 Heat transfer analysis of blast furnace stave *International Journal of Heat and Mass Transfer* **51(11–12)** 2824–33

[3] Wu LJ, Gao X J and Wang S 2013 Global optimization of blast furnace cooling stave based on grey correlation analysis *Journal of Tongji University(Natural Science)* **41(12)** 1885–88

[4] Guo G S, Zhang J L, Li F G, et a1 2015 Research on structure optimization of cast iron cooling stave with elliptic water pipes *Energy for Metallurgical Industry* **34(4)** 28-32

[5] Liu Q and Cheng S S 2016 Heat transfer and thermal deformation analyses of a copper stave used in the belly and lower shaft area of a blast furnace *International Journal of Thermal Sciences* **100** 202-12

[6] Guo Z Y, Liang X G, Zhu H Y 2006 Entransy——a physical quantity to describe the heat transfer ability of an object *Progress in Natural Science* **16(10)** 1288–96

[7] Han G Z, Guo Z Y 2006 Two different thermal optimization objective functions: dissipation of heat transport potential capacity and entropy production *Journal of Engineering Thermophysics* **27(5)** 811–13

[8] Zhu H Y, Chen Z J and Guo ZY 2007 Experimental study on the electro-thermal simulation of the principle of extremum entransy dissipation *Progress in Natural Science* **17(12)** 1692–98.

[9] Liu X B, Guo Z Y and Meng J A 2008 Entransy analysis of thermal resistance in heat exchangers based on entransy dissipation *Progress in Natural Science* **18(10)** 1186–90

[10] Qian X D and Li Z X 2011 Analysis of entransy dissipation in heat exchangers *International Journal of Thermal Sciences* **50(4)** 608–14

[11] Feng H J, Chen LG, Xie Z H, et a1 2014 Constructal entransy dissipation rate minimization for gas-turbine blade cooling *Journal of Mechanical Engineering* **50(4)** 142–49

[12] Xu Y C, Chen Q and Guo Z Y 2015 Entransy dissipation-based constraint for optimization of heat exchanger networks in thermal systems *Energy* **86** 696-708

[13] Guo J F and Huai X L 2016 The heat transfer mechanism study of three-tank latent heat storage system based on entransy theory *International Journal of Heat and Mass Transfer* **97** 191-200

[14] Cheng X G, Meng J A, Guo Z Y 2005 Potential capacity dissipation minimization and entropy generation minimization in heat conduction optimization *Journal of Engineering Thermophysics* **26(6)** 1034–1036

[15] Zhang Z M, Ren Z P and Mei F M 2007 *Heat Transfer* (Beijing: China Architecture and Building Press) 153–54

[16] Heinrich W G, Nikolas S and Wolfram G 1992 Die physikalischen verhaltnisse im bereich der kohasiven zone des hochofens (teil 1: grundlagen und modelle) *Stahl und Eisen* **112(8)** 73–79

Acknowledgments

This research subject is funded by the National Natural Science Foundation of China (51574179)

IOP Conf. Series: Journal of Physics: Conf. Series **1072** (2018) 012019 doi:10.1088/1742-6596/1072/1/012019

Health, Environmental Effects, the Control of Emission from Power Plants and the Need for a New Emission Capture Technology.

D K De[1, 2], U I Ikono[1] and S O Akinmeji[1]

[1]Department of Physics, Covenant University, Ota, Nigeria
[2]Sustainable Green Power Technologies, Mansfield, Texas

Email: dilip.de@covenantuniversity.edu.ng; dlpd770@gmail.com

Abstract. It is well known that global warming is a major environmental concern and it is primarily caused by the release of pollutants (greenhouse gases) into the atmosphere. Electric power plants have been found to be major contributors of environmental pollution which results in health and environmental hazards. There have been several ongoing research on ways to curtail and capture emissions which have led to development of state-of-the-art technologies. However, these technologies have not been proven cost effective and as a result, they cannot be implemented on a global scale. Also, chemical based flue gas emission capture technology may lead to secondary pollutions. Prevailing cryogenic capture technologies show inadequacies due to cost ineffectiveness and its high energy demand. This paper focuses briefly on the effects of emissions from power plants on health and environment and talks about the need of a new single technology that captures each component of flue gas emissions without the use of chemical reagents and offers an affordable cost for global implementation for clean energy and environment.

1. Introduction

With increase in greenhouse gas emission, there is need to harmonize technological advancements with safe environmental practices. Pollutants have great influence on environmental quality and health on local, regional and global scales. However, current control measures for cutting down the continual emission rate are limited by exorbitant cost of implementation, operation and maintenance.

1.1. Health and environmental effects of emissions from coal and natural gas power plants.

The combustion of coal and natural gas in power plants [1] generates a number gases which are released into the atmosphere. They include: carbon dioxide (CO_2), steam, particulate matter, nitrogen oxides (NO_x) (x= 0.5, 1, 2, 3), sulfur oxide (SO_2), sulfur trioxide (SO_3), sorbent injection, ash, soot, mercury (Hg, Hg^{++}, Hg) and carbon monoxide. Sulfur oxide emission from power plants and cements factories poses various health and environmental threats which include: skin irritation, breathing difficulties, chronic bronchitis, eye irritation with corneal haze, pulmonary edema, psychic alterations, severe irritation from noise, throat, lungs and airway inflammation, mucus secretion, ozone –layer depletion, deforestation, corrosion of building materials and paints, acidification of waterways to the detriment of aquatic life. Both sulfur oxide (SO_2), sulfur trioxide (SO_3) form sulfuric acid droplets by reaction with atmospheric oxygen and moisture. These can form suspension of fine particles known as sulfate aerosols, which can alter the microphysical and optical properties of clouds and this plays a major role in the uncertainty of cloud models. Nitrogen dioxide (NO_2) emission results in formation of secondary pollutants such as ground level ozone/petrochemical smog, particulate matter and constituents of acid rain. It forms secondary particles called nitrates which causes haze and reduces visibility through the absorption of full spectrum of light. Exposure to nitrogen dioxide affects the human respiratory tract and increases vulnerability to, respiratory infections, coughing, wheezing, asthma, and lung cancer. Reacting with atmospheric oxygen and moisture, NO_2 yields nitric acid particles which can remain suspended in air and form aerosols. Both types of aerosols reflect sunlight and reduce visibility and cause a cooling affect. N_2O has a global warming potential of 298 – 310 [2] and atmospheric life time of 114 years. It leads to the formation of acid rain and is also a greenhouse gas that causes the depletion of the ozone layer. Mercury in flue gas from power plants can undergo reactions to transform it to extremely toxic compounds such as methylmercury (CH_3Hg) or dimethyl mercury CH_3HgCH_3 which are deposited on lakes, trees, and can be consumed by birds and fish [3]. Oxides of mercury are soluble in water, thus they can be absorbed by fish. Approximately 75 tons of mercury is emitted from coal power plants in the USA yearly, even after the removal of 30 tons of mercury by washing the coal before it is burnt. Exposure to mercury emissions results in mercury poisoning which has profound cellular, cardiovascular, hematological, pulmonary, renal, immunological, neurological, endocrine, reproductive and embryonic toxicological effects. Particulates pose various adverse effects on health and environment such as volcanic eruption due to changes in earth's climate, drought, rainfall decline, global dimming, ocean acidification, asthma, lung cancer, cardiovascular disease, premature delivery, birth defects, vascular inflammation, atherosclerosis, etc. The earth's climate system is undergoing aberrations as a result of carbon dioxide, methane, soot, nitrous oxide and other greenhouse gases being released into the atmosphere. They act as a blanket, trapping the heat from the sun and causing it to warm the planet. Some of the environmental and health effects of global warming are: melting of snow and glacier, drought, rise in sea level and coastal flood, heat waves, deforestation, increasing smog, asthma outbreak, malaria due to mosquito increase, presence of new pest, extinction of habitats such as coral reefs and alpine meadows, shrinking water supplies, changes in food supply, increasing effects of severe weather etc. Exposure of the human body to too much CO_2 creates acid imbalance in the lungs as there is more carbon dioxide absorbed in proportion to oxygen. This affects the internal respiration and the victim may experience symptoms such as asphyxiation, frostbite, kidney damage, coma, headaches, dizziness, nausea, increased heart rate, cardiac arrhythmia, memory disturbance, lack of concentration, restlessness, sweating, vomiting, eye and extremity twitching, visual and hearing disturbances (including hearing loss and ringing in the ears, photophobia, blurred vision, transient blindness), convulsion and death. According to a research paper by Mark Jacobson of Stanford University, anthropogenic carbon dioxide emission contributes not only to climate changes but also leads to about a thousand deaths and many more. Cases of respiratory illness and asthma in the United States per year with per degree Celsius rise in temperature is associated with air pollution. Globally the figure is 20,000 deaths per year, per degree Celsius rise. However, according to him the total global annual deaths due to environmental pollution is around 2 Million. World Health Organization (WHO; 2007) specified the premature death caused by CO_2 globally as 2 million (China -65000; India – 530,000; USA – 41,000)]. The high cost of current technologies has limited the implementation of current technologies for controlling toxic gases from power plants and capturing industrial CO_2 emissions especially in developing countries.

Existing clean energy technologies for removal and control of toxic emissions from coal combustion.

Research on pollution and emissions from power plants has led to the development of technologies used in capturing and reducing emissions. Some of the current technologies for control of pollution (except carbon dioxide and carbon monoxide) used particularly in coal power plants include: (i) electrostatic precipitators (ESP), (ii) fabric filters, (iii) mechanical collectors, (iv) venturi gas scrubbers, (v) selective fuel switching, (vi) selective non catalytic reduction (SNCR) and selective catalytic reduction (SCR), (vii) wet flue gas desulfurization, (viii) dry FGD utilizing spray dryer absorber (SDA) [4], (ix) circulating dry scrubber (CDS), (x) dry sorbent injection (DSI), (xi) activated carbon injection technique, (xii) enhanced wet technology, (xiii) K-fuel technology. In this paper discussions are made on the costs of implementing and operating some of these technologies. We may briefly describe them, and discuss also, the need for a cost effective technology that can capture all the components of flue gas.

1.2. Cost involved in using Flue Gas Desulfurization (FGD) systems for SO_2 removal.
Typically, the estimated cost of building a new coal power plant is £800million, the installation of FGD systems incurs an additional cost. Between the years 2008- 2010, installation cost of FGD equipment was estimated to be $150/kW (that is $150 million for a 1000MW coal power plant.) which is about 11% the cost of setting up the plant itself. Also, the FGD system is found to cause a reduction in the net power generation efficiency of about 2 to 4% due to increased parasitic load. It was predicted in 2009 that at the end of the next 12 years, about $200 billion will be spent in introducing FGD systems to new and existing combustion units [5]. SO_2 removal cost which depends on the sulphur content of the fuel, power plant capacity, etc. ranges from $200-$2000 per ton. Using scrubbers, it costs about $50,000/MW for 70% removal and $250,000 for 95% removal [3]. The costs of wet and spray FGD systems for various power plant capacities [6] are displayed in table 1 [7] below.

Table 1: Summary of FGD cost information in $/MW (2001 Dollars)

Scrubber type	Unit Size (MW)	Capital Cost ($/kW)	O & M Cost ($/kW)	Annual Cost ($/kW)	Cost per ton of Pollutant removed ($/kW)
Wet	>400	100-250	2-8	20-50	200-500
	<400	250-1,500	8-20	50-200	500-5000
Spray	>200	40-150	4-10	20-50	150-300
	<200	150-1,500	10-300	50-500	500-4000

*(EIA, 2002; EPA, 2002; Srivastava 2001 [7].

From the above table1, the cost of using FGD is high for smaller power plants.

1.3. Costs involved with chemical method for control of $\boldsymbol{NO_X}$ *using chemicals/ reagents.*
The installation cost of SCR is approximately $200 per kW and NO_x removal cost per ton ranges from $1,400 to $3,000. This is dependent on the power plants size [8,9]. Reports from air pollution control technology factsheet estimate the cost of NO_x removal per ton to be:
i. Industrial coal boiler: $2,000 - $5,000
ii. Industrial oil, gas and wood fired boiler alongside an electrostatic precipitator for particulate matter: $1,000 - $3,000
iii. Large gas turbine: $3,000 - $6,000
iv. Small gas turbine: $2,000 - $10,000.
As a result of application of modern emission controlling technologies for SO_2 and NO_x the emissions are gradually decreasing in USA but not in India as can be seen in the data provided in the Appendix.

1.4. Cost of mercury removal.
Using ACI, the US Department of Energy estimated the cost of mercury removal to range from $$37,500-445,000/lb. mercury [3] and the cost of using carbon injection to range from $25,000-$70,000/lb. mercury extracted. The EPA estimated the cost of 90% removal using carbon injection to range from $5000- $28000/lb. mercury [10]. All of which are quite expensive. Also, it is said that these removal costs can increase subsequently the price of electricity (per kWh) by 0.1 to 0.8 cents [11-12], It is also noteworthy that in the ACI, the fly ash will contain carbon and mercury (making the fly ash unsalable) which is very hazardous and its disposal incurs additional costs to electricity

generation. However the figures listed above did not include disposal costs nor loss of income from the sale of fly ash.
Current techniques of capturing Carbon dioxide

1.5. Post-combustion.
This process involves the cooling of emitted CO_2 gas from power plants and industries, the cooled gas is afterwards bubbled through a vertical column absorber packed with solvents such as ammonia or amine which can preferentially absorb the CO_2 (nevertheless, researchers are still in search of more efficient solvents). The amine solution (MEA) is dropped from the top of the vertical column (in some systems) and the reaction takes place on the beads, the reaction between the MEA(R-NH2) and the CO_2 in the flue gas forms a weakly bonded but stable compound called carbamate. Challenges arising from this technique includes amine losses due to thermal or oxidation degradation. Oxidation degradation can be controlled using oxygen inhibitors however thermal degradation is still a cause for concern. It occurs predominantly during the regeneration reaction where high temperature is experienced in the packing column [13-14] (this gives an explanation for 20-30% of all amine losses) [15]. Also, degradation products such as nitrosamines and nitramines which are highly soluble in water [16] are shown to be geno-toxic and possible carcinogens [17]. Available data on the 23 nitrosamines that may form during amine based capture has been studied and 2 out of the 23 have been declared probably carcinogenic to humans by the International Agency for Research on Cancer, while the 7 are classified as possibly carcinogenic to humans. Not much data is available on nitramines but they are likewise suspected to be carcinogens. The presence of oxygen or oxidizing gases in flue gas can corrode amine systems. Corrosion rates can be reduced by employing corrosion inhibitors, lowering the concentration of MEA, using appropriate materials for the construction of the amine system, and using temperature controls [18]. All of which contribute additional costs to the amine capture system.

1.5.1. Installation cost for amine technology.
The cost of scrubbing CO_2 is 25-40% as much as the total output cost of electricity which is very expensive; especially for older plants operating at 35% efficiency or lower [19]. Husbye et al [19-21], presents a cost analysis of amine based CO_2 capture, according to him, it costs about 750 euros per ton for a flue gas with 2.5% CO_2 concentration, while for 20.5% CO_2 concentration; it decreases to 375 euros per ton. Even for a 90-100% capture from industrial plants, the cost is still quite high. The power plant design by Choi et al shows that the installation cost for a plant with total removal capacity of 2820 tons of CO_2 per day is $114 million [20]. Also, the carbon capture and storage (CCS) plant at Mongstad in Norway has an investment cost of NOK 20-25 billion [22] which is equivalent to $2.4 to $3 billion.

1.6. Metal Organic Frame (MOF) Work Technology.
It involves filling a large column with MOF balls and passing flue gas through one end of the column and receiving the flue gas with significantly less CO_2 out of the other end. The process is repeated a few times on the exit gas stream in order to ensure maximum adsorption of CO_2. Thereafter, the CO_2 is heated to a climacteric temperature to release the CO_2 which is obtained in a fairly pure form. The released CO_2 is afterwards compressed and transported to its required destination. Almost 220MW power has to be diverted for the capturing of CO_2 using MOF column in a 500MW plant, leaving only 280-300MW at most for use. It costs about $2.9 billion to fill the columns with MOF balls and these costs more than 1000 times the cost of current amine based technology. Asides the capital cost (which is very high), the operational costs is about 3 times the cost of the most cost effective amine based technique [23]. This technique is still in its research stage and has not been commercialized. At this current stage of development, it is also not practicable in developing countries. Other technology applied in cutting down emissions includes carbon sequestration, the use of oxyfuels, breaking down of lignite coal to syngas which is then used as a fuel in power generation, the Climeworks capture device which is designed to capture CO2 directly from the air (costs about $1000 per ton of CO2), etc.

All of these have their individual economic and operational concerns. Analysis by Jeffrey Rissman and Robbie Orvis [24] shows that coal plants equipped with current CCS are almost three times more expensive than onshore wind power and more than twice as expensive as solar photovoltaics (PV). Though these costs will decline with continuous research and development, the potential for cost improvement is limited. Coal with CCS will always need significant subsidies to compete economically with wind and solar.

Global costs of carbon capture and storage.

The capital cost of the CCS facilities, even with latest technological advancements is still in the order of billions of dollars- consequently it is not affordable and practicable on a large scale in many countries yet (especially, developing countries like India). Kemper County employed new coal-gasification technology which is reported to have been commissioned after a total expenditure of about $US7.5 billion. The initial cost of the 500 MW power plant project was $1.8 Billion. After which the coal giant Southern Company (SC) announced in June 29, 2017 its plans to discontinue work on the carbon capture segment of its power plant in Kemper, Mississippi. SC has decided not to use the amine based technology to capture the plant's greenhouse gas emissions. Due to the addition of CCS facility to the Boundary Dam, the planned cost shot up terrifically, however, the high costs were attributed to many other factors aside CCS. The world's second largest "Petra Nova CCS Project" located in Thompsons, southwest of Houston, Texas commenced in 2010. This was preceded by the announcements made by the U.S. DOE. The project commenced at a cost of $1billion. Although these projects have made somewhat progress in capturing CO_2 emissions, the impact is very minute when viewed on a global scale and these technologies cannot be implemented worldwide due to their heavy costs. Therefore, there is a need for better and cost efficient technology that can capture components of emissions, particularly CO_2 and store the capture product in a cost effective manner.

Need for a new emission capture technology.

Due to continuous increase in the demand for energy, and the inability of renewable energy technologies to effectively replace the use fossil fuels fully, the world will continue to face increasing release of emissions into the environment, unless controls are put in place to combat it. Emissions such as mercury has the ability to travel long distances thus its effect can be experienced on a global scale. In order to curtail the effects of these emissions (such as a global warming, which requires more than 50% carbon capture to have significant impact), the need for emission capture technologies cannot be overemphasized. However, as discussed earlier, current emission control technologies are very expensive are not practicable in developing countries. Some technologies like the amine based carbon capture result in harmful secondary pollution e.g. nitrosamines and nitramines which are found to be possible carcinogens. Therefore, there is need for a new single technology to capture each component of the emissions cost effectively, without use of any chemical reagent. The technology should be such that a portion of the captured components would pay for the capture cost and ensure that the storage of the captured components is much easier and cost effective, compared to the current methods of transporting captured CO_2 through pipe lines to empty underground coal or oil field. If such a technology is effectuated in power industries, it will eliminate pollution and its resulting adverse effects on health and environment. In the second article "*PATHWAY(S) TO CLEAN AIR, MITIGATION OF CLIMATE CHANGE AND 100% RENEWABLE/CLEAN ENERGY TRANSFORMATION*" [25] we shall briefly discuss such technology, which has been researched in depth theoretically and discovered to be the most cost effective, and energy efficient technology that can achieve 100% capture of emission in a way that each component can find industrial use. The procedure is based on a novel cryogenic capture technique using innovative ideas such as generating auxiliary power using the heat of the flue gas. In this new technology the nitrogen, which constitutes nearly 70% of the flue gas is cooled to a few degrees above its boiling point. This is used to fractionally cool the incoming flue gas, after being subject to specified pressures and the ashes, oxides of mercury etc. are separated by using existing technologies and special ceramic filters. The flue gas is expanded at several stages and cooled further. The process separates individual component of flu gas into liquid form while the CO_2 is separated in the form of dry ice. The new technology is being patented by the Dilip K. De (Provisional US patent (# 62593828 filed on Dec. 31, 2017. Non-provisional patent #15915007, dt. March 7, 2018)). Only electrical energy with fixed amount of water is used in this capture technology. The net energy cost estimate for capturing 1 ton of CO_2 is about \$8 per ton, if auxiliary power is generated using the heat of the flue gas and about \$25 per ton, if auxiliary power is not generated. However, these cost estimates include capture of all the affiliated pollutants along with each ton of CO_2 capture. Thus the technology will be much more cost effective than any existing technology.

Appendix

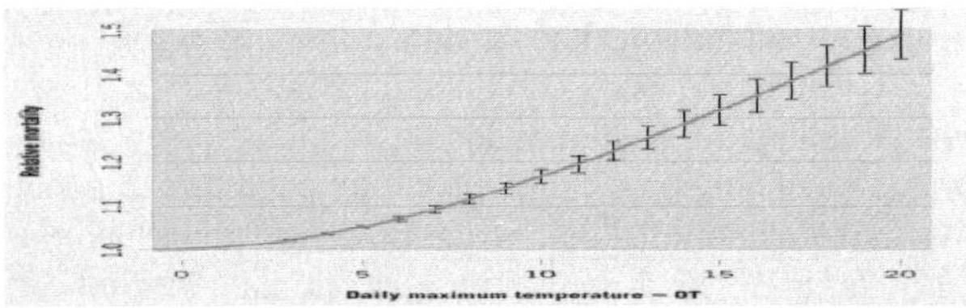

[a]Graph shows the Cumulative effect over 15 days. The analysis was produced using results from Honda et al. (2014) [27]

Figure A1 shows Correlation[a] between daily maximum temperature minus the optimum temperature, and relative mortality for people aged over 65 years. [26]

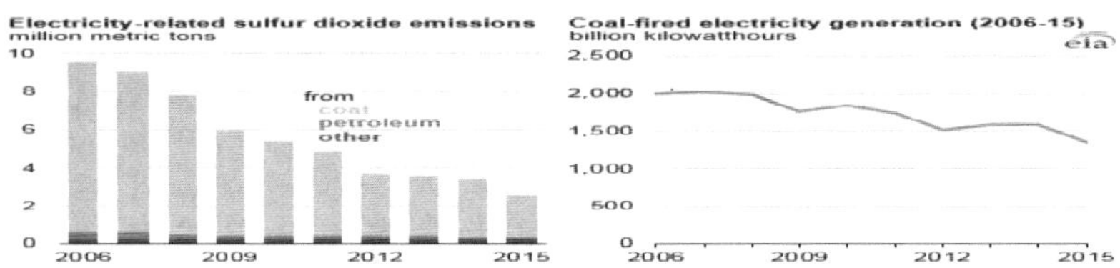

Fig. A2 [28] Shows the sulphur dioxide emissions in USA from coal fired electric power plants during the period 2006-2015. It shows that the emission declined gradually over years. It is primarily due to expensive emission control technologies put in some of the plants. However, these technologies cannot be implemented on a global scale due to heavy installation and operation costs.

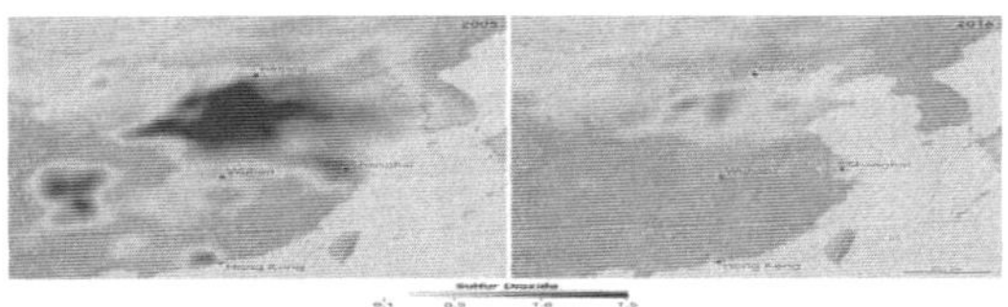

Fig. A3 [29]
The satellite data from NASA shows significant decline of SO_2 in 2016 china, compared to that in 2005 in China. This can be attributed to the emission control technologies in China.

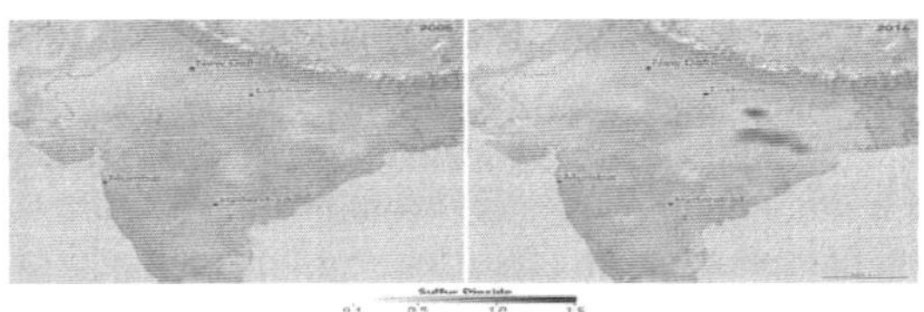

Fig. A4 [29] NASA satellite data shows significant growth of SO_2 emissions from power plants in India in 2016 relative to that in 2005. Available emission control technologies cannot be afforded in developing such as India

Table A1 [30] Nitrogen oxide emissions in six Midwest states in 1999 (tons per year)

NO_x Emissions	All Sources	Power Plants	Power Plant %	Coal Plants	Coal Plants as % of Power Plants
Illionois	1,112	279	25.1	268.4	96.2
Indiana	860.3	350	40.7	349	99.7
Michigan	859.9	207	24	183.6	88.7
Minnesota	487.6	87.2	17.9	83.8	96.1
Ohio	1,149	431.4	37.5	429.5	99.5
Wisconsin	506.8	109.6	21.6	106.5	97.2
Total	4,975.6	1454.3	29.4	1420.9	97

The table A1 above shows that coal power plants are key contributors in the release of NO_x emissions into the environment.

References

[1] L Moretti and CS Jones. *Advanced emissions control technologies for coal-fired power plants-technical paper*, BR-1886-Babcock & Wilcox, Power Generation Group, Inc., Barberton, Ohio, U.S.A., Presented to: Power-Gen Asia, Date: October 3-5, 2012,Location:Bangkok, Thailand.

[2] U.S. Environmental Protection Agency, Office of Atmospheric Programs. *N_2O-greenhouse gases and global warming potential values excerpt; inventory of U.S. greenhouse gas emissions and sinks: 1990 - 2000*, EPA 430-R-02- 003, April 2002

[3] Brian H Bowen, Marty W Irwin. *Basic mercury data & coal fired power plants CCTR Indiana centre for coal technology research. March 2007 CCTR basic facts file #2* https://www.purdue.edu/discoverypark/energy/assets/pdfs/cctr/outreach/Basics2-Mercury-Mar07.pdf

[4] Paul S. Nolan, Babcock & Wilcox Company Barberton, Ohio, USA. *Flue gas desulfurization technologies for coal-fired power plants*. Presented by Michael X. Jiang, Coal-Tech 2000 International Conference. November 13-14, 2000 Jakarta, Indonesia

[5] Zifeng Lu and David G Streets. *Increase in NO_x emissions from Indian thermal power plants during 1996–2010*. Unit-based inventories and multi-satellite observations, environ. Sci. technology, 2012, 46 (14), pp 7463–70, DOI: 10.1021/es300831w

[6] Air Pollution Control Technology fact Sheet- EPA-452/F-03- 034 (https://www3.epa.gov/ttncatc1/dir1/ffdg.pdf).Srivastava Ravi K, US Environmental Protection Agency, Office of Research and Development. EPA 2000. *Controlling SO_2 emissions: a review of technologies*. EPA/600/R-00/093, Washington, D. C. 2002. http://www.epa.gov/ordntmt/ORD/WebPubs/SO2/Index.html.

[7] Srivastava R K and W Josewicz. *Flue gas desulfurization: the state of the art*. Air and Waste management Assoc. 51:1676-1688, 2001. IA.2002: EIA-Database. *Annual steam-electric plant operation and design data, 2001, energy information administration*. Department of Energy, November 26, 2002. http://www.eia.doe.gov/eneaf:/electricity/page/eia767.html.

[8] J Edward Cichanowicz. *Current capital cost and cost-effectiveness of power plant emissions, control technologies*. John L Sorrels, David D Randall, Karen S Schaffner and Carrie Richardson Fry. *Selective catalytic reduction* http://www.publicpower.org/files/pdfs/uargscr_fgdfinal.pdf

[9] K Bruce, C Castaldi, J Cook and D Lachapelle. *EPA region 3 low-NO_x, control technology study*. Environmental Corporation, Acurex Report FR pp 97-116

[10] *Hg removal cost estimation*. https://hub.globalccsinstitute.com/publications/coal-quality-impacts-and-gas-quality-control-oxy-fuel-technology-carbon-capture-and-storage-cost-impacts-and-coal-value/62-hg-removal-cost-estimation

[11] Pavlish, J H. *Fuel processing technology* 2003 pp 82, 89-165.

[12] Daniel C. Mussatti, Dr. Ravi Srivastava, Paula M. Hemmer and Randy Strait. *Selective catalytic reduction*. EPA/452/B-02-001. https://www3.epa.gov/ttn/catc/dir1/cs4-2ch2.pdf. October 2000.

[13] Eide-Haugmo I, Lepaumier H, Einbu A, Vernstad K, da Silva E F and Svendsen H F (2011) *Chemical stability and biodegradability of new solvents for CO_2 capture*. Energy Procedia 4, pp 1631-36.

[14] Boot-Handford M E et al (2014). *Carbon capture and storage update*. 2014. Energy and Environmental Science 7, pp 130-189

[15] Shao R and Stangeland A (2009). *Amines used in CO2 Capture –health and environmental impacts*. The Bellona Foundation. The Bellona Report: www.bellona.org

[16] DaSilva E, Booth A, Loeppky R, Ma'Mun S, Matthias K, Heimstad E, Johnson M and Feilberg K. *Protocol for evaluation of solvents – process and atmospheric chemistry*. Sintef report, 19 June, 2010.

[17] Låg M, Lindeman B, Instanes C, Brunborg G and Schwarze P. *Health effects of amines and their derivatives associated with CO_2 capture.* The Norwegian Institute of Public Health, 2011 ISBN 978-82- 8082-462-2

[18] Barchas, R. and R. Davis (1992). *The Kerr-McGee/ ABB Lummus Crest technology for the recovery of CO_2 from stack gases.* Energy Conversion and Management, 33(5-8), pp 333-340

[19] Jo Husebye, Amy L Brunsvold , Simon Roussanaly , and Xiangping Zhang. *Techno economic evaluation of amine based CO2 capture: impact of CO_2 concentration and steam supply.* Energy Procedia 23 (2012) pp. 381 – 390.

[20] Gerald N Choi et al https://www.netl.doe.gov/publications/proceedings/04/carbon-seq/154.pdf

[21] Anand B. Rao. *Details of a technical, economic and environmental assessment of amine-based CO_2 capture technology for power plant greenhouse gas control.* Report Submitted October, 2002, Work performed under Contract No.: DE-FC26-00NT40935, https://www.cmu.edu/epp/iecm/IECM_Publications/2001rb%20Rao%20&%20Rubin,%20DOE%20Progress%20Append%20(Oct)-%

[22] Ravnum S, Rundén-Pran E, Fjellsbø LM and Dusinska M. *Human health risk assessment of nitrosamines and nitramines for potential application in CO_2 capture.* Regul Toxicol Pharmacol. July, 2014; 69(2):250-5. doi: 10.1016/j.yrtph.2014.04.002. Epub 2014 Apr 18. https://www.cmu.edu/epp/iecm/IECM_Publications/2001rb%20Rao%20%26%20Rubin%2C%20DOE%20Progress%20Append%20(Oct)-%20Rev%20June%202012.pdf

[23] Furukawa H et al. *Ultrahigh porosity in metal-organic frameworks.* Science. 2010 Jul 23;329(5990):424-8. doi: 10.1126/science. 1192160. Epub 2010 Jul 1.

[24] Jeffrey Rissman and Robbie Orvis. *Carbon capture and storage: An expensive option for reducing US CO_2 emissions.* https://www.forbes.com/sites/energyinnovation/2017/05/03/carbon-capture-and-storage-an-expensive-option-for-reducing-u-s-co2-emissions/#6e8af15d6482.

[25] Submitted to: Renewable Transformation Challenge 2017 https://www.elsevier.com/physical-sciences/energy/renewable-transformation-challenge#Winners .

[26] WHO Library Cataloguing-in-Publication Data Quantitative risk assessment of the effects of climate change on selected causes of death, 2030s and 2050s.*World Health Organization.* ISBN 978 92 4 150769 1 (NLM classification: WA 30.5)

[27] Honda Y et al. Heat-related mortality risk model for climate change impact projection. *Environ Health Prev Med.* 19 pp. 56–63, (2014).

[28] EPA reports. https://www.eia.gov/todayinenergy/detail.php?id=29812

[29] Irene Ying. *China's sulphur dioxide emissions drop, India's grow over last decade.* 13 November, 2017. https://www.nasa.gov/feature/goddard/2017/chinas-sulfur-dioxide-emissions-drop-indias-grow-over-last-decade

[30] Ellen Baum, David Schoengold, Amy Hennen, Izaak Walton and Kurt Waltzer.Ohio Environmental Council Bruce Hill. Clean air task force. http://www.catf.us/resources/publications/files/Midwest_Nitrogen.pdf

The Research of Theoretical Approaches to Increasing the Reliability of Automated Control Systems in the Energy Sector

Kyaw Zin Lin[1], Aung KyawMyo[2], Portnov Evgeni Mikalovic[3] and KyawZawYe[4]

[1,2,3] Department of Informatics and Computer Software System, National Research University of Electronic Technology, Bld.1, Shokin Square, Zelenograd, Moscow, 124498, RUSSIA

[4] Department of electronic, telecommunication system and Automatic control system, Pyinoolwin Research Center, Republic of the Union of Myanmar

[1] kyawzinlinn51@gmail.com, [2] akyawmyo54@gmail.com, [3] evgen_uis@mail.ru, [4] kyawzawye85@gmail.com

Abstract. The article deals with the problems of transferring reliable signals of control and control over the communication channels of complex power supply management systems for production facilities. The discrepancy between the concept of "information reliability" and "information security from interference in the communication channel" is mathematically justified. It is theoretically justified that in order to ensure a high level of the system reliability of equipment state signals in the process of technical diagnostics, it is necessary to combine the procedures for inputting information from sensors with encoding. A method of three-pulse correlation coding based on the coding procedure for all sensors simultaneously in three time-separated stages has been developed. It forms two allowed and four forbidden codes that determine the presence and the type of malfunction or the absence of distortions in the information message. A technique for calculating the system reliability of the signals of the state of electrical equipment characterized by the sum of the probabilities of undetectable distortions of information and undetectable faults of the elements of the entire information delivery route and the component $1{,}1\cdot10^{-12}$ has been proposed. This is several orders of magnitude better than the parameters of modern analogues.

1. Introduction

In accordance with GOST 26.205-88, the reliability of the automated control systems in the energy sector is estimated in size of the probability of undetectable distortion of information with the channel interference. According to this definition, reliability is the degree of conformity between the received and transmitted signal, in this case the channel interference is a distorting factor causing the nonconformity.

So, in the technical documentation on control systems and technical diagnostics of various producers it is shown that a high level of reliability (probability of undetectable distortion is $\sim10^{-10}\ldots10^{-16}$)of the received information is achieved through the use of powerful anti-interference codes oriented to the protection against the distortion of messages with the channel interference. At the same time, this indicator does not take into account the possibility of distortion of the information before its receipt in the encoder, therefore a high level of reliability is provided only at one of the sections of the information delivery route from the sensor to the receiver [1].

Hence, a fuzzy definition of the basic characteristic of control systems of the reliability of information has led to a lack of interconnection between the coding and decoding methods from the principles of construction of technical diagnostic systems and the architecture of functional modules.

In this interpretation of the reliability the malfunction of the equipment systems which do not lead to any failure of the command or signal transmission and to undetectable distortion control commands or status signals of the equipment are not taken into consideration although it is obvious, that failures to execute commands are incomparably less likely to lead to emergencies than direct executing a false command. It is necessary to emphasize that this nonconformity of responsibility is reflected in GOST 26.205-88 [2]. Especially, the standard levels of allowable failure probability from executing a false control command or receiving false equipment state signals with undetected distortions is by several orders of magnitude lower than the levels of probability of the distorted data. Therefore, it is obvious that to determine the level of reliability of information in the systems of technical diagnostics, for example, control commands is necessary in accordance with the probability of false execution of the command control without specifying the cause and location of the command distortion. It is obvious that the reliability of information and other channels of technical diagnostics systems should be evaluated in this way[2].

If we analyze the most popular control systems and technical diagnostics applying the proposed principle, it turns out that having a very low probability of undetectable distortion of commands, signals and measurements with the channel interference they have much lower real reliability of the received information due to the lack of interconnection between the principles of building coding and information transfer modules.

2. Methods of information transmission, coding methods and architecture of functional devices

In order to correlate the probabilistic measure of technical diagnostics system which are determined by the methods of information transmission, coding methods and architecture of functional devices, we introduce the indicator of transformation of reliability of the received information determined by the probability ratio of the undetected distortion of commands, signals or measurements of the interference in the communication channel to the probability of a false command or receipt of false signals and measurements regardless of the cause and place of distortion[3].

The probability of undetectable distortion of information signals and control commands with the channel interference - P_{unde_SC} is determined by the power of the applied code. Knowing the average rate of the flow of information transfer μ_{avg}, the average expected time between failures - T_{fail} can be determine.

$$T_{fail} = \frac{P_{unde\_SC}}{\mu_{avg}}. \tag{1}$$

Taking the flow of non-detectable failures as poisson we will receive:

$$\mu_{avg} = \frac{P_{unde\_SC}}{T_{fail}} = \frac{P_{unde\_SC} n_{req}}{t_h} [hour^{-1}], \tag{2}$$

Where n_{claims} - the average number of claims per year,

t_h - the number of hours per year.

We divide the system hardware into n groups according to the potential fault tolerance criterion. The intensity of undetected failure of the element in the equipment of the $i-th$ group is denoted as μ_i, and the probability of reception and transmission of the distorted information in the event of undetected failure - as P_i. Then the degree of transformation of the reliability of the received information can be defined as:

$$\Omega = \frac{\mu_{avg} + \sum_{i=1}^{n} P_i \mu_i}{\mu_{avg}}. \tag{3}$$

To illustrate the possible values of coefficient **Ω**, we substitute the numerical values of the parameters into the formula (3).

Let's take the probability of non-detectable distortion of information signals due to interference in the communication channel equaled to $P_{unde\_SC} = 10^{-12}$ (in accordance with GOST 26.205-83) and provided through the use of error-correcting code, the average number of requirements for switching the state of the object for a year $- T_{requ} = 10^3\, year^{-1}$.

Suppose that, only one element is installed in the transmission of state signals by this object (the switch is in the input circuit), the failure of which is not detected by monitoring and diagnostic devices[4,5]. The standard failure intensity of this element is $\mu_i = 5 \cdot 10^{-7}\, hour^{-1}$ and the probability of the occurrence of the signal distortion state because of the fault in the given element will be taken as $P_j = 10^{-4}$. Consider that there are no other sources of distortion of information signals from formula (1), we obtain the following equation:

$$\mu_{avg} = \frac{10^3 \cdot 10^{-12}}{9 \cdot 10^3} = 1{,}1 \cdot 10^{-13}\, hour^{-1}$$

Then the degree of transformation of reliability will be:

$$\Omega = \frac{1{,}1 \cdot 10^{-13} + 5 \cdot 10^{-7} \cdot 10^{-4}}{1{,}1 \cdot 10^{-13}} = 0{,}5 \cdot 10^3$$

It is apparent that even in this assumption there are sufficient "soft" exposure conditions of influence on the system equipment, while real indicators of the reliability of equipment state signals are found to be several orders of magnitude worse than expected.

To find the ways to improve the reliability of technical diagnostics systems, consider the standard route of passing signals from equipment status sensors to the display elements which are shown in figure 1.

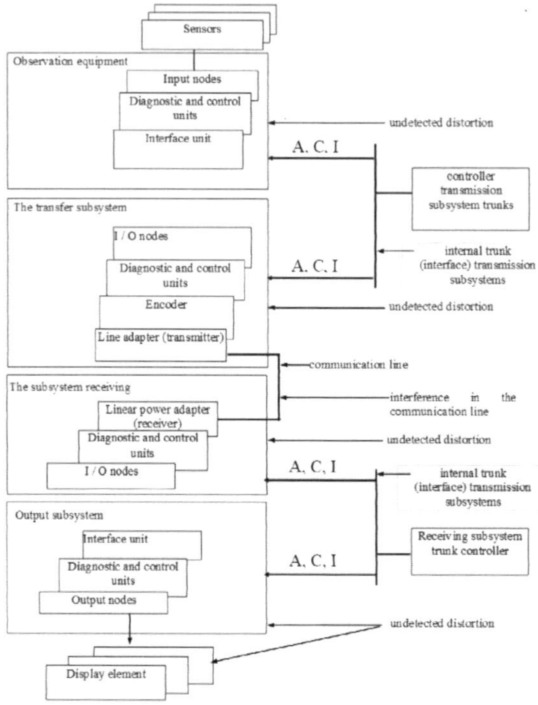

Figure 1. The model of the part of technical diagnostics system

The data from the input subsystem hardware status signal source with the permission from the controller of the internal trunk of the subsystem passes through the interface on which the signals of address (A), control (C) and information (I) are sent to the data transmission subsystem in the communication channel.

The factors determining the level of reliability of the information transmitted on the internal highway are:

- failures and malfunctions of hardware and software of subsystems of input and information transfer,
- interference received in the input / output circuit of interface signals.

The information from sensors in the area from data input to the transmission subsystem encoder is exposed to interfering factors - interference ("tips" on the communication circuits of the module with sensors), faults and failures. These "disturbing" factors can lead to detectable and undetectable distortions.

With the help of “powerful” anti-interference codes one can detect almost all possible channel interference.

However, the encoder-decoder pair shown in figure 1 cannot ensure the data reliability, in case of receipt of information in the encoder with previously undetected distortions. Thus, the "standard" principles of building of the devices (with emphasis on information protection only against distortion with the channel interference) do not guarantee the reliable reception of data from sensors. Moreover, to the mentioned above possible reasons for entering false information the following should be added:

- undetectable distortion of data in the internal trunk of the receiving subsystem;
- undetectable distortions due to interference, equipment failures and malfunctions in the data output area of the source module.

It should be emphasized, that the impact of" interfering " factors outside the communication line of the control station in modern devices are increased for a number of reasons:

- the transition to the use of microprocessor elements in which the sensitivity to external action is incomparably higher than that of relay or semiconductor elements,
- Increase in the speed of information input from the sensors and reduction of the energy ratio of operating signals and interference,

- the transition to digital communication channels in which the impact of interference is lower than that of traditional analog, for example, high frequency communication channels.

It is obvious that taking into account the above factors to improve the actual reliability it is necessary to take additional measures.

3. The channels of information feedback (IF) and decision feedback (DF)

Usually, in addition to the use of powerful anti-interference codes, feedback channels are used to improve the reliability of information transmission. Currently, the most common of these are the channels of information feedback (IF) and decision feedback (DF), the scheme of which is presented in Figures 2 and 3.

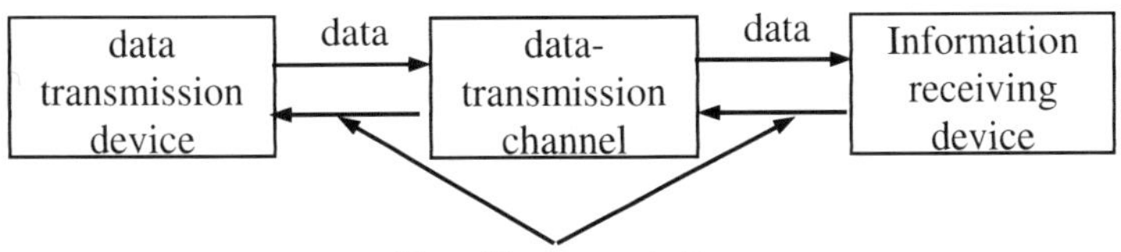

Figure 2. System with decision feedback

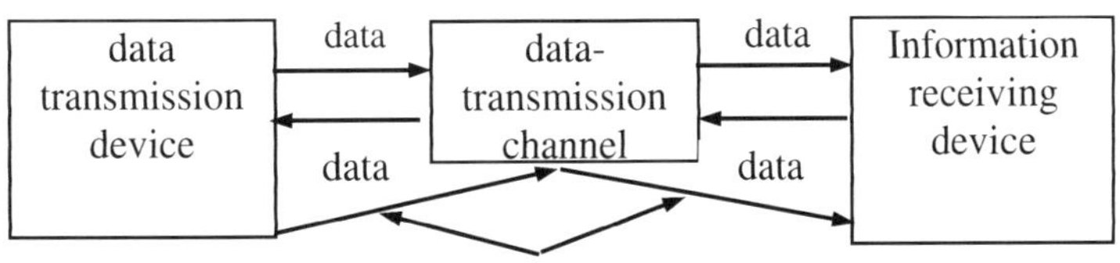

Figure 3. Systems with information feedback

In systems with decisive feedback, the receiver makes a decision on the absence or presence of errors/distortions based on the analysis of the received data. In the first case, "receipt" is transmitted in the direction of the transmitter, confirming the undistorted reception of information. In systems with informational feedback, the receiver returns a received message to the transmitter. The decision on the absence of distortion is made by the transmitter, which transmits to the receiver a "receipt", according to which the receiver realizes the previously received information.

The probability of distortion of data in the channels with DF is higher, since the control is carried out only on the message received from the transmitter, and in the channel with IF the control is carried out both on "direct" and on "return" message. However, if a "direct" message is distorted, even a system with information feedback does not provide a blockage of distortions.

Therefore, traditionally used methods for constructing technical diagnostic systems with common encoder and decoder for all modules do not provide reliable data reception from sensors.

According to the results of the analysis the following conclusions can be drawn:

- to improve the reliability of the information, it is necessary to enter the diagnostic nodes that allow in a dynamic mode to detect distortions along the entire route of information delivery to the receiver,
- The time shift between failure detection and the display of it at the central control point of the system must be minimal,
- the diagnostic nodes should provide a dynamic control of the operability of all elements of the system, that is to detect faults equivalent to short circuits ("excess" signals "1") and breaks of connections (("excess" signals "0").

Thus, in order to ensure a high level of system reliability of equipment state signals in the process of technical diagnostics, it is necessary to combine the procedures for inputting information from sensors with coding, that is to include the signal coding device in the information input module. Dynamic

performance monitoring can be successful if during the input of information combined with its coding, one can check the possibility of installing any module element in the state "1" and "0".
Similar to the traditional double-pulse signal coding Manchester code is used [7], in which each discrete signal is represented by two bits "10" or "01", the second bit being formed by simply inverting the first. This code allows to detect the distortion of information message signals with the channel interference of the controlled point in which an encoder with a central control point (information decoder) is installed.
But the presented coding method based on the use of Manchester code does not allow to detect the distortions caused by a rupture or a short in the encoder's communication circuit with any sensor, that is it does not provide the required level of reliability.
To eliminate this disadvantage, a coding method, based on the coding procedure was developed for all sensors simultaneously in three time-separated stages-at the first stage for each sensor of the discrete signal, a signal "1" or "0" is formed for the closed and open state respectively, at the second stage determine the presence of a short in the circuit or breakage of the sensor communication circuits with the encoder is determine, and at the final third stage the serviceability of the output code generation elements is identified.
It should be noted that the three-stage coding procedure is carried out in such a way that the combination of the signals obtained at all three stages when coding the signal from the sensor is not repeated in the closed and open state of the sensor, the presence of a short in the circuit and a break in the sensor communication circuit with the encoder, and also at malfunction of the encoder and (or) the controller of data transfer to the communication channel.
Simultaneous implementation of coding procedures for all sensors provides the increased efficiency and the use of three stages of coding allows to create the non-repeatable code combinations allowing to identify the presence and type of malfunction of the elements participating in coding, forming and transmitting of the generated message through the communication channel with the central control point.

4. Method of three-pulse correlation coding based on the coding procedure

Figure 4 show the code combinations of the three-pulse correlation code (TCC) formed at three time-separated stages when coding signals from sensors and when generating the message transmitted over a communication channel.
In the closed state of any sensor, code combination 110 is formed at three stages of coding and when the open state – code combination is 010. If the open circuit fault of sensor is fixed with the encoder device, at three coding stages, code combination 011 is formed for the corresponding sensor, and when detecting of short circuit in the circuit connection of the sensor with encoder - the code combination is 101. In case of failure type" false operation " of elements, that forming an information message for transmission over the communication channel code combination 111 is formed, and in the event of a failure type "failure" - the code combination is 000. Thus, a unique code combination is formed for each considered case, in the analysis of which by the presence and type of failure or absence of distortion in the information message received at the central control point is definitely determined.

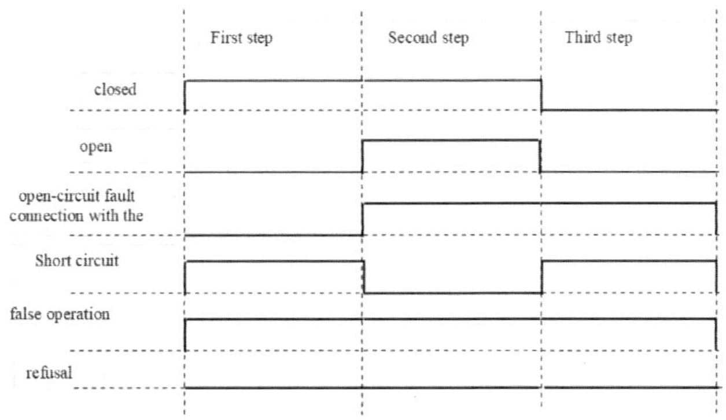

Figure 4. The code combinations of three-pulse correlation code

In the figure 5 has shown the implementation of coding procedures in dividing each of the three stages into three clock period. The total number of clock period, in which three stages of coding are realized, is equal to nine (from zero to eight). The number of clock period in each stage is equal to three. On the first clock period of each stage of coding (1-1; 1-2; 1-3) the state of sensors is fixed, and the received information is written to the memory register. Simultaneously data is read from the random access memory (RAM) with writing data to the register, which is received on earlier procedures of coding of signals from sensors.

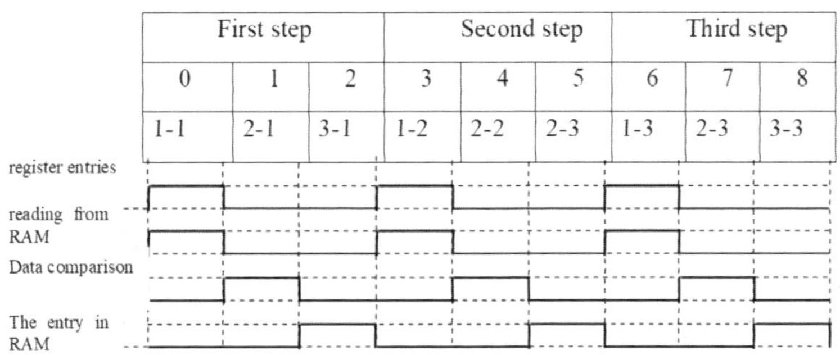

Figure 5. The implementation of coding procedures

At the second clock period (2-1, 2-2, 2-3) of each coding step the data written to the register and corresponding to the current state of each sensor and its communication circuits are compared with the information read from the RAM. Nonconformity of the compared signals indicates about changing the status of any sensor or its circuits. At detection a nonconformity for any sensor, new information on the third clock period of the corresponding stage (3-1; 3-2; 3-3) is entered into RAM and also into the transfer controller. The controller generates the information message for transmission over the communication channel in the form of a cyclic code containing except the status codes of sensors, communication circuits with them and information about the performance of the equipment, additional components- the address of the controlled point, the code of the information type as well as the sequence of control code (SCC).

Let's calculate the system reliability of electrical equipment status signals for the proposed method of coding.

The calculation will be carried out under the following initial conditions which determine the structure of the information package:

d_1 -identification code of the controlled point address (two bytes);

d_2- code identifying the type of information messages (two bytes);

d_3 – information field (0 to 64 bytes);

d_4 - the field of protection, control sequence code (two bytes).
Provided that the distortion of each component of the information message is an independent event determine the probability of undetectable distortion of the entire message as the sum of the probabilities of all its components:

$$P_{unde} = P_{input} + P_{cpa} + P_{type} + P_{prote} \tag{4}$$

Where P_{input} – undetectable distortion when entering and encoding simultaneously for the formation of a three-pulse correlation code;
P_{cpa} –probability of undetectable distortion of the identification code of the controlled point address;
P_{type}–probability of undetectable distortion of message type identification code;
P_{prote} – the probability of undetectable distortion of the code control sequence.
The probability of undetectable distortion of information when entering information and forming a three-pulse correlation code is:

$$P_{input} = nP_0 P_{gat}^2 (1 - P_{gat})^{n-2} \tag{5}$$

Where P_0 - the probability of such interference impact in the case of the repeated distortion of the input signal of the state of equipment which is opposite to the effect of the primary distortion;
n - bit code combination equal to the number of sensors ,

$$P_{gat}^2 = \left(P_{dis} \frac{T_{gat}}{T_{cycle}} \right)^2 , \tag{6}$$

Where P_{dist} - the probability of a single signal distortion due to interference;
T_{gat} - signal duration gating signal from the sensor;
T_{cycl} - the period between adjacent cycles of survey of the sensor status
In the equation (5), the factor n is used, but not the number of triple combinations of distortions, so as to three-shots of the correlation code pairs of the control signals is carried out separately.
At $(1 - P_{gat})^{n-2} \to 1$ will get;

$$P_{input} = nP_0 \left(P_{dis} \frac{T_{gat}}{T_{cycle}} \right)^2 \tag{7}$$

Using the check sequence cyclic code provides the receipt of a code distanced$d \geq 4$.
Therefore, when applying the control sequence to the input P_{input} , we obtain

$$P_{input} = \left(nP_0 \left(P_{dis} \frac{T_{gat}}{T_{cycle}} \right)^2 \right)^4 C_n^4 . \tag{8}$$

Taking into account the length of the identification codes of the KP address, the type of information message and the control sequence of the cyclic code, we obtain an indicator

$$P_{cpa} + P_{type} + P_{prote} = P_{dis}^4 \left(C_{d_3}^{2d_1} + C_{d_3}^{2d_2} + C_{d_3}^{2d_4} \right) \tag{9}$$

Here d_3 is determined based on the bit width of the information for each sensor – 6 bits (3 bits of forward and inverse code) and the total number of sensors equal to $n = 32$ .Then $d_3 = 6n/8 = 24\ byte$
Consequently:

$$P_{unde} = \left(nP_0 \left(P_{dis} \frac{T_{gat}}{T_{cycle}} \right)^2 \right)^4 C_n^4 + P_{dis}^4 \left(C_{d_3}^{2d_1} + C_{d_3}^{2d_2} + C_{d_3}^{2d_4} \right) \tag{10}$$

Substituting the numerical values: $P_{dis} = 10^{-4}$, $n = 32$, $T_{gat} = 10^{-7} c$, $T_{cycle} = 10^{-2} c$, $P_0 = 1/8$, $d_1 = d_2 = d_2 = 2$ will obtain $P_{unde} \approx 1{,}1 \cdot 10^{-12}$.

The resulting value satisfies the most stringent requirements of the standard and it is almost completely determined by the probability of distortion of additional components of the information message which makes it more relevant to use more secure identification codes of the controlled point address and message type.

When assessing the reliability index, it should be taken into account that all modules of the information delivery route from the source module to the receiver are practically simple re-transmitters and do not affect to the format of the information message.

Thus, the system reliability of the information P_{syst} can be represented by the sum P_{unde} and the sum of the probabilities of undetectable faults of all modules of the information delivery route from the source module to the receiver P_{fault} :

$$P_{syst} = P_{unde} + P_{fault} = \left(nP_0 \left(P_{dis} \frac{T_{gat}}{T_{cycle}} \right)^2 \right)^4 C_n^4 + P_{dis}^4 \left(C_{d_3}^{2d_1} + C_{d_3}^{2d_2} + C_{d_3}^{2d_4} \right) + \sum_1^m \left(P_i \right)^4 \cdot C_{q_i}^4 \tag{11}$$

Where P_i -the probability of occurrence of a single fault element of the $i = th$ module,

$C_{q_i}^4$ - the number of combinations of four elements of the $i = th$ module,

q_i – the number of elements in the $i = th$ module.

Indicator "4" takes into account the minimum code distance of the information message to be generated.

By accepting $P_i = 10^{-6}, q_i = 0.5 \cdot 10^3, i = 5$ obtain

$$P_{syst} = P_{unde} + P_{fault} = 1{,}103 \cdot 10^{-12} \tag{12}$$

Thus, the proposed coding principle provides a high level of system reliability of information.

5. Conclusion

A new approach to improving the reliability of automated control systems in the power industry has been proposed in this study. The approach is based on the generalization of the most important information characteristics of the system into a single indicator-speed, noise immunity, reliability, validity. System reliability is determined through the probability of undetectable information distortion, regardless of the cause and the location of distortion along the source-receiver path, provided that the time lag between the time of occurrence and undistorted reception of the message is less than the limit established as a criterion for converting the received message into an unreliable category. The basis of the new theory is the principle of coding which is based on the conversion of the Manchester code into conditionally correlated one with the automatic insertion into the transmitted message of the usability condition of thc communication circuits of each sensor with the encoder. It provides the diagnosis of the reliability of information along the entire route of its delivery from the sensor to the receiver. It is mathematically justified that the probability of undetectable distortion of the information provided by the proposed encoding method is 3-4 orders of magnitude better than that of known analogs. The proposed approaches will form the basis for the creation of noise-immune encoding for commands of control of power units.

6. References

[1] Ye, K.Z., Portnov, E.M., Gagarina, L.G., Lin, K.Z. Method for increasing reliability for transmission state of power equipment energy//*2015 IEEE Global Conference on Signal and Information Processing, 14-16 Dec. 2015 – Orlando, FL*, 2016. P. 433-437

[2] Evgeni M. Portnov, Kyaw Zin Lin, Vitaliy V. Kolkin, Aung Kyaw Myo. Method for Detection of Failures in Complex Process Control Systems// *2017 IVth International Conference on Engineering and Telecommunication (EnT) (IEEE), 29-30 Nov. 2017, Moscow, Russia.*

[3] Kyaw Zin Lin, Portnov E.M. Development of methods for increasing the efficiency of information exchange in control systems of power energy// *ISER – 83rd International Conference on Science, Technology, Engineering and Management (ICSTEM) Sydney, Australia,* December 5th -6th 2016

[4] Black Yu Computer networks: protocols, standards, interfeysy.-M .: Mir, 1990.-p 506 c.

[5] A. R. Di Fazio • T. Erseghe • E. Ghiani •M. Murroni • P. Siano • F. Silvestro// Integration of renewable energy sources, energy storage systems, and electrical vehicles with smart power distribution networks, Received: 25 January 2013 / Accepted: 9 April 2013 / Published online: 8 May 2013_ *Springer-Verlag Berlin Heidelberg 2013.*

[6] Candy, R Changing the telecontrol protocol to support substation based intelligent alarm processing and direct control staff interaction with the substation RTU / gateway / / *IEEE PES Powe-rAfrica 2007 Conference and Exposition, PowerAfrica; Johannesburg; South Africa*; 16 July 2007 through 20 July 2007

[7] Carré, O., Englert, H., Ostermeier, A., Lange, J., Lhuillier, P. Substation to control centre communication based on IEC 61850: Requirements, concepts and practical experiences / / *44th In-ternational Conference on Large High Voltage Electric Systems 2012; Paris; France*; 26 August 2012 through 31 August 2012

[8] Han, G., Xu, B. IEC 61850-based open communication system of advanced distribution automation Dianwang Jishu / *Power System Technology Volume 35, Issue 4, April 2011*, P 183-186

[9] Li, L., Miao, P., Wang, J. Analyze and implementation of substation automation system base on IEC 61850 standard (2006) *Power System Technology, 30 (S1),* p 321-324.

[10] Xu, B., Li, T. *Investigations to some distribution automation issues (2010)* Dianli Xitong Zi-donghua/Automation of Electric Power Systems, 34 (9), p 81-86.

[11] Li, M.-Z., Li, Y.-S., Xie, Y.-K. Failure analysis and maintenance of automatic communica-tion channel (2009) Dianli Xitong Baohu yu Kongzhi / *Power System Protection and Control*, 37 (23), p 167-169.

Acknowledgement

The authors are grateful to head of department of Informatics and Computer Software Systems professor Gagarina L.G. and professor Lisov O.I. for help in writing an article.

IOP Conf. Series: Journal of Physics: Conf. Series **1072** (2018) 012021 doi:10.1088/1742-6596/1072/1/012021

Comparison of the Multi-Directional Delivery Efficiency of Low-Tonnage LNG and Pipeline Gas in Russia

Pavel Tcvetkov, Daniil Pritulyak, Dmitriy Tananykhin

Saint-Petersburg mining university, 2, 21st Line, St Petersburg 199106, Russia, pscvetkov@yandex.ru

Abstract. One of the fastest growing areas of the gas industry is low-tonnage LNG production, which allows to improve the efficiency of the gas distribution process due to the flexibility of the supply routes. A small number of articles devoted to this field, indicate that LNG is less attractive than pipeline gas from an economic point of view, however, such estimates do not take into account the flexibility of LNG delivery routes. This article proposes a model for a comparative assessment of the LNG and pipeline gas price. The distinctive feature of the model is the possibility of changing the number of delivery directions, which makes it possible to evaluate the impact of the LNG transportation flexibility. The results show that in contrast to the pipeline gas, the increasing in the number of supply routes with the same total volume of deliveries, does not reflect on the price of LNG. In addition, this causes a decrease in the minimum delivery distance at which LNG automobile transportation can compete with a pipeline.

1. Introduction

At present, natural gas is one of the most demanded types of energy resources, despite the intensive development of renewable energy. According to significant number of forecasts, its position in the 21st century will only increase because of the cost, energy and environmental characteristics [1].

In this regard, the issues of increasing the efficiency of gas use and reducing its price for consumers are becoming more urgent. These issues are also significant for Russia, which ranks second in terms of proven reserves of natural gas, as well as second in terms of its production. In 2016, only PJSC Gazprom supplied more than 356.6 billion cubic meters of natural gas to 20 countries [2].

Transportation of natural gas is carried out, as a rule, through a system of gas pipelines [3], which is associated with a number of territorial, political and economic problems. Firstly, it is necessary to take into account the peculiarities of the terrain, along which the pipeline is laid. Secondly, carrying out a gas pipeline through the territory of other countries requires a consent of local authorities. Thirdly, delivery through the pipeline is absolutely inflexible and does not allow to change the direction of gas supplies in a short time. Fourthly, the gas pipeline system requires constant technical support throughout its entire length. Nevertheless, this is the most frequently used method of gas delivery to a significant part of the global market [4].

An alternative to pipeline transport is the use of liquefied natural gas (LNG) technological chains, which became a point of interest in recent decades with the desire to solve the problems of gas pipelines operation. Large and medium volumes of LNG over a long distance as a rule shipped by a Maritime transport [5]. In the case of low-tonnage LNG production, the delivery can be carried out by automobile, rail and river transport. Given that transport networks are developed much better than pipelines, this allows diversifying the supply routes.

In major part of research papers, the comparison of LNG and pipeline gas prices is carried out on the basis of gasification projects consideration for one specific point of gas supply. However, this approach does not allow estimating the effect of the technological schemes flexibility. In this regard, the aim of this paper is to assess the feasibility of low-tonnage LNG multi-directional delivery based on a comparison of its unit price with the price of pipeline gas.

2. Methodology

The assessment was conducted for Russian economy conditions, where are significant problems with gas supply in the far regions. According to some estimates, between 40% and 50% of the country's territory does not have access to centralized energy supply, with the major part of these locations in the Eastern part of Russia [6]. Despite this, the annual gas consumption in the domestic market is more than 450 billion cubic meters, and the forecasted growth is 1-3% per year.

In general, there are more than 18,000 small settlements on the territory of Russia, about 30-40% of which are not gasified [7]. The construction of gas pipelines in these settlements cannot be met without significant state support, due to the capital intensity of such projects, and, consequently, significant unit costs at low volumes of gas consumption.

2.1. Formatting the title

The strategy of Russian territory gasification is based on the development of the main gas pipeline, the continuation of which for Siberia territory is planned only in a long term. Partially the problem of far regions gasification could be solved by the development of the LNG industry [8].

In addition to the objective need of liquefaction, the LNG production process is associated with substantial expenses for regasification [9, 10]. However, when we talk about small volumes of consumption, it is possible to use modern mobile power plants that can function on LNG due to the regasification systems included in their structure [11]. It allows to reduce production costs significantly, as well as increase the attractiveness of the trigeneration technological schemes [12]. The cost of such facilities operating on LNG is slightly higher (10-15%) than the equipment operating on a pipeline gas. Therefore, conducted assessment did not take into account the difference in the costs associated with energy generating facilities.

Transportation of low-tonnage LNG is possible by automobile, rail and river transport. However, the railway network is not sufficiently developed in regions that do not have access to centralized energy supply [13]. The same drawback is typical for river shipping, and in this connection, attention in this article was focused on automobile delivery.

The idea behind this article is that the regional gasification project with using LNG will be considered more efficient if the increase in the unit price of LNG (in terms of natural gas) after automobile transportation from the point of production to the points of consumption will be lower than with using a pipeline for the same distance. At the same time, distance delivery means that there is at least one point of consumption within the radius of this distance; however, it must be taken into account that the number of such points may vary significantly (Figure 1).

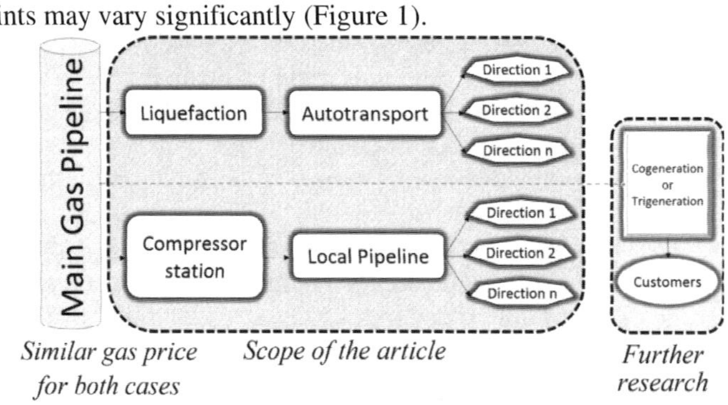

Figure 1. Scope of the article

For the purpose of the assessment and to show the dynamics of pipeline gas price changing, the maximum number of delivery directions accepted equal to 5. The assessment was carried out for LNG

IOP Conf. Series: Journal of Physics: Conf. Series **1072** (2018) 012021 doi:10.1088/1742-6596/1072/1/012021

plant with a capacity of 22,500 tons of LNG per year, which is equal to 31.056 million cubic meters of natural gas. In original, all calculations were made in Russian Rubles. The conversion was carried out at the rate of 58 RUR per 1 USD.

2.2. Liquefaction plant CAPEX and OPEX

The structure of the plant construction CAPEX is shown in Figure 2-B. The depreciation rate is 10%. OPEX of the LNG plant consist of the purchased natural gas price, the salary of employees, the cost of electricity and unforeseen expenses. The structure of the OPEX for the production of one ton of LNG is shown in Figure 2-A.

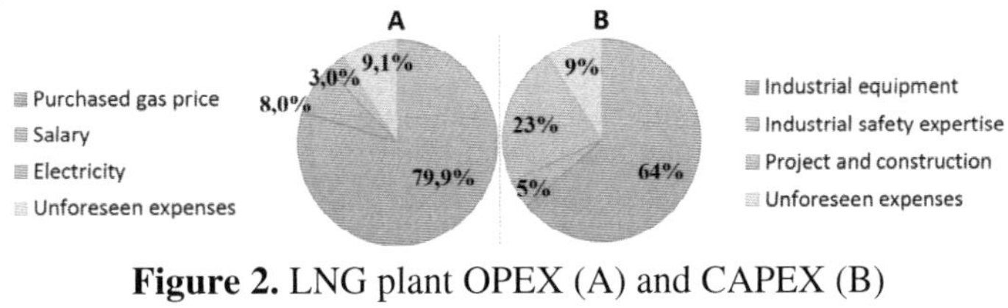

Figure 2. LNG plant OPEX (A) and CAPEX (B)

2.3. Automobile delivery CAPEX and OPEX

CAPEX of automobile delivery are limited by the purchase of trailers with tanks. Based on the analysis of acceptable vehicles, trailers at a price of 224 137.93 USD, which are capable to deliver 19 tons of LNG per voyage were chosen.

Automobile delivery OPEX consist of the fuel price (0.655 USD per liter), drivers' wages, maintenance costs and unforeseen expenses.

The wage of drivers is equal to the average market rate for Russia - 4.268 USD per hour, including all taxes and fees. At the same time, depending on the transportation distance, increasing coefficient was introduced, according to table 1.

Table 1. Increasing coefficient for drivers' wages

Indicator	Values					
Distance [km/month]	<100	100-500	500-1000	1000-3000	3000-6000	>6000
Coefficient	1	1.1	1.2	1.3	1.4	1.5

In addition to the basic 5%, the amount of unforeseen expenses takes into account the risks arising with the increase in the transportation distance, such as breakage, traffic accidents and jams etc. An increase in the basic level of unforeseen expenses occurs every 500 km by 1%. For example, for a distance between 1000 to 1500 km, unforeseen expenses will be 7% of the total transport OPEX.

Calculation of the LNG unit price, taking into account transportation by trailers, was carried out according to the following formula:

$$P_{LNG} = P_{NG} + \frac{P_L}{k_1} + (N_t \cdot P_t \cdot DR_t + Ex_{OP}^t) / Vol_{ng} \tag{1}$$

where P_{NG} - price of purchased natural gas from the main pipeline, RUR/m3; P_L- price of liquefaction per ton of LNG, RUR/t; k_1 - coefficient to convert LNG volume to natural gas (1.38026); N_t - number of trailers; P_t - price of one trailer, RUR; DR_t - depreciation rate of trailers; Vol_{ng} - total volume of LNG in terms of natural gas, m3; Ex^t_{OP} - transport OPEX (formula 2).

$$Ex_{op}^t = N_t \cdot N_v \cdot l \cdot P_G \cdot V_t \cdot L + N_d \cdot h \cdot S_d \cdot k_2 + UE_t \tag{2}$$

where N_v - number of voyages per one trailer; l - the distance between LNG plant and point of destination, km; P_G - price of gasoline, RUR/liter; V_t - LNG tank capacity, tons; L - load rate, %; N_d - number of drivers, people; h - working hours in estimated period, hour; S_d - wage of a driver, RUR/hour; k_2 - coefficient to increase a wage (table 1); UE_t - unforeseen expenses, RUR.

2.4. Pipeline CAPEX and OPEX

The capacity of a gas pipeline depends on its diameter and the working gas pressure. The determination of the required maximum capacity is based on the need to transport 31,056 million m3 of natural gas (22 500 tons of LNG) per year. Approximate calculations to select the required diameter of a pipe can be performed with a high degree of certainty by formula 3.

$$D = \sqrt{\frac{Q_{\max}}{0.67 \cdot (p_{work} + 1.033)}} \tag{3}$$

where Q_{max} – maximum capacity of the gas pipeline, m3/h; p_{work} – working gas pressure, MPa.

Gas pipelines construction CAPEX include the cost of compressor stations (CS), the cost of pipes, the cost of laying the pipeline and unforeseen expenses. For the assessment, five possible pipeline diameters were selected, depending on the number of delivery directions. The volume of supplies in each direction decreases in proportion to their total number. Table 2 shows the initial data for analysis.

Gas pipeline OPEX are mainly related to the need to maintain the operation of compressor stations (95% OPEX). Based on the analysis of the existing compressor stations, taking into account the required volume of supplies, the following was accepted: for 90 mm diameter, OPEX of one compressor station is 3034.5 th. USD per year; for 62 mm - 2827 USD / year; for 51 mm - 2621 USD per year; for 45 mm - 2414 USD per year; for 41 mm - 2206 USD per year.

Table 2. Basic technical parameters for the pipeline gas price calculation

Indicator	Units	Number of delivery directions				
		1	2	3	4	5
Diameter	Mm	90	62	51	45	41
Capacity	th.m^3/h	3,8	1,8	1,2	1,15	0,96
Work pressure	MPa	0,6	0,6	0,6	0,75	0,75
Distance between CS	Km			160		
Price of CS	th. USD	20.7	13,8	10,3	8,6	5,2
Price of the pipe	th. USD/km	5	2.9	2.6	1.7	1.6
Pipeline laying	th. USD/km	18,9		17.2		
Unforeseen expenses	%			10		
Depreciation rate	%			4		
OPEX	th.USD/km-year	5.2	3.5	2.4	1.2	0.7

The cost of pipeline gas was calculated with using the following formula:

$$P_{PG} = n \cdot (Ex^{pipe} + Ex^{cs}) / Vol_{ng} \tag{4}$$

where *n* - number of delivery directions, according to table 2; Ex^{pipe} - expenses for a pipeline (formula 5), RUR; Ex^{CS} - expenses for a CS (formula 6), RUR.

$$Ex^{pipe} = l \cdot (1 + UE_{pipe}) \cdot (Ex_{op}^{pipe} + DR_{pipe} \cdot P_{pipe}) \tag{5}$$

where UE_{pipe} - unforeseen expenses for the construction and operation of the gas pipeline, unit fraction; EX_{op}^{pipe} - pipeline OPEX, RUR/km per year; DR_{pipe} - depreciation rate of a pipeline, unit fraction; P_{pipe} - price of the pipe, according to table 2.

$$Ex^{cs} = N_{cs} \cdot (1 + UE_{cs}) \cdot (Ex_{op}^{CS} + DR_{cs} \cdot P_{cs}) \tag{6}$$

where N_{cs} - number of CS; UE_{cs} - unforeseen expenses associated with CS, RUR; Ex_{op}^{cs} - CS OPEX, RUR/year; DR_{cs} - depreciation rate of the CS, unit fraction; P_{cs} - price of the CS, RUR.

2.5. Proposed technical-economic model

To carry out the assessment, a technical-economic model was developed (Figure 3). This model allows to compare two options for region territory gasification, taking into account the required volume of gas supplies and the number of supply routes from the “starting point”. The “starting point” is a place located near the main gas pipeline, where it is possible to build an LNG plant, or to create a branch of a low-capacity gas pipeline to supply several locations with gas. The key technical and economic parameters specified in the model were discussed above.

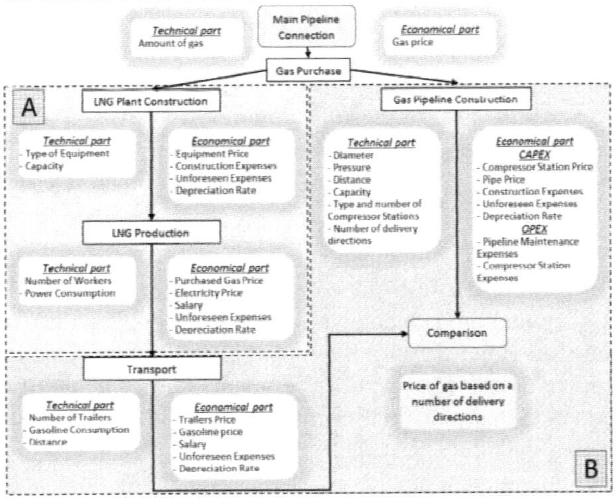

Figure 3. Technical-economic model

The proposed model could be divided into 2 parts: Part A. The economic evaluation was carried out on the basis of similar projects analysis. The result of this evaluation is plant’s OPEX (201 USD per ton of LNG) and CAPEX (13.4 mln USD). Part B. Calculations were carried out on the basis of a mathematical model expressed by the formula 7.

$$\begin{aligned} & f(n,l) \rightarrow f(n_{\max}, l_{\max}) \\ & Min(P_{LNG}, P_{PG}) \\ & \begin{cases} 0 < l \le l_{\max} \\ N_t = N_d \\ L = 1 \\ N_v, N_t, N_{cs}, Vol_{ng} > 0 \end{cases} \end{aligned} \tag{7}$$

2.6. Assumptions and simplifications

Calculations are based on a comparison of the LNG price (P_{LNG}), expressed in a similar volume of natural gas in cubic meter, with the price of natural gas transported by a pipeline (P_{PG}). The delivery distance (l) varies from 1 km to l_{max}, with l_{max}=7500 km. The value l_{max} is the maximum theoretical distance that one trailer can pass and go back to the plant in one month.

To conduct a comparative assessment of transportation to several locations, the total volume of transported gas was divided by the number of directions (n). At the same time, the delivery distance was accepted equal for all directions. For example, transportation in two points at a distance of 1000 km. means that 11 250 tons of LNG will be delivered to each point and the distance from the plant to each point is 1000 km.

N_t=N_d indicates that the number of drivers is equal to the number of trailers, since an excessive increase in the number of employees will lead to a significant increase in total wages, especially with long distance delivery. On the other hand, the lack of drivers will not allow to transport all produced LNG.

L=1 indicates that the loading of each tank is 100%. In other words, for one trip a trailer carries 19 tons of LNG. Thu number of voyages (N_v), the number of trailers (N_t), the number of CS (N_{ks}) and the volume of purchased natural gas (Vol_{ng}) must be greater than zero to exclude errors in the calculations.

3. Results

Calculations have shown that the attractiveness of using LNG produced at low-tonnage plants instead of pipeline gas depends on the range and number of supply directions (Figure 4).

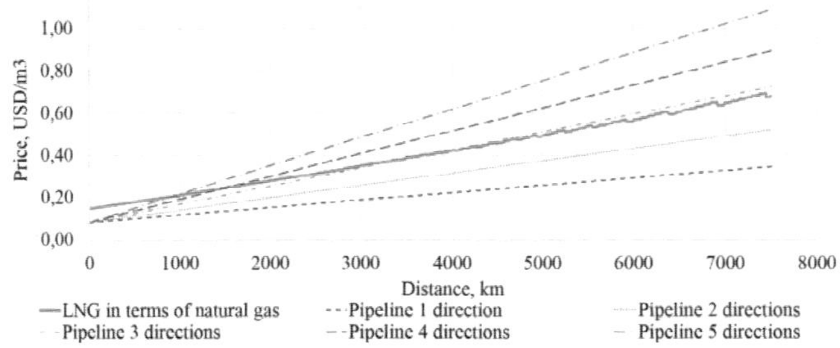

Figure 4. Results of gas price modeling

In general, the result showing that the use of LNG in 1 and 2 directions of supply is more expensive than pipeline gas, does not contradict world practice and is one of the most significant reasons for the predominance of a pipeline transport. However, if we talk about 3 or more directions of supply, which is possible, for example, in the implementation of the regional gas supply strategy, LNG can be considered as a competitor to a pipeline gas (Figure 5).

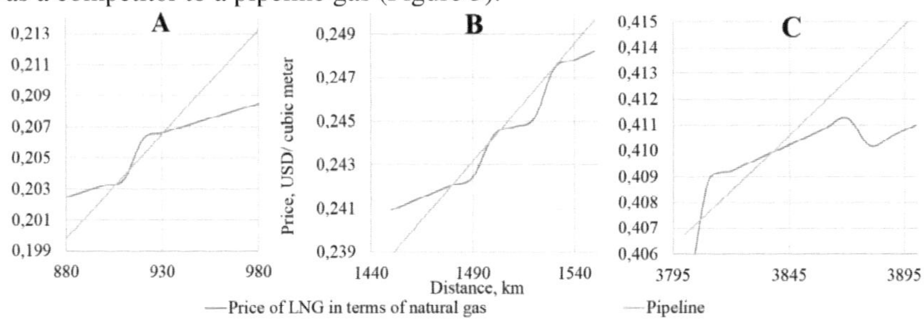

Figure 5. Multi-directional delivery distances with equivalent prices

For delivery in 3 directions (Figure 5C) LNG can compete with natural gas at a distance of over 3 845 km. However, the theoretical possibility of such long-distance delivery by trailers does not necessarily mean its implementation. For example, in Russia, such a range of deliveries is possible for transporting LNG from the European part of the country to Siberia, but in most cases, the location of the main gas pipeline allows the use of pipeline gas at a much shorter transportation distance.

However, for delivery in 4 (Figure 5B) and 5 (Figure 5A) directions LNG can compete with pipeline gas at much smaller distances. Given the vast territory of Russia and a large number of small settlements, it is already possible to talk about the possibility of improving the strategic energy supply programs of the regions by using low-tonnage LNG plants. In general, it can be expected that further increase in the number of supply directions, an increase in the efficiency of the gasification process and a decrease in the cost of trailers will lead to increasing in LNG attractiveness at distances of less than 500 km.

4. Conclusion

At present, the development of LNG industry is constrained by high production and transportation costs compared to pipeline gas. This fact is confirmed by significant number of research papers devoted to the evaluation of medium - and large-tonnage LNG production projects, as well as by small number of studies of low-tonnage production, which in most cases do not allow to conduct a fair comparison with pipeline gas. The technological schemes with use of LNG allow to provide a greater flexibility of supply, which makes them more attractive than pipeline transport for supplying several remoted locations.

The relevance of the development of low-tonnage LNG production for the Russian market lies in the possibility of creating schemes for supplying energy resources to sparsely populated areas and remote regions, in which the gas pipeline is unprofitable and inexpedient in a view of the relatively small

volumes of demand for energy raw materials. However, despite the results obtained during the evaluation, for the large-scale implementation of the LNG projects, it is necessary to introduce additional measures of state support to compensate a part of production and transport costs.
It can be argued that LNG production plants have good prospects for development in Russia and abroad. However, there are too few research devoted to the solution of problems of low-tonnage LNG production and the prospects of their large-scale implementation. In this regard, further work will be aimed at improving the accuracy and complexity of the model for assessing LNG projects, both in terms of production and transport solutions, and in the use of LNG or regasified gas. This can be of significant practical importance in view of the reorientation of energy markets from petroleum raw materials to natural gas.

5. References

[1] Makarov AA. et al. World energy forecast until 2040 and consequences for Russia, Moscow, 2013.
[2] Cherepovitsyn AE., Kraslawski A. Research into the innovative potential of an oil and gas company at different stages of field development. Journal of Mining Institute, 2016, 222, 892-902
[3] Shangaraeva L., Tananykhin D. The Development of Energy-Saving Technologies of Oil Wells Chemical Treatment to Prevent Scale. International Journal of Applied Engineering Research, 2015, 10–24, 44157-44161.
[4] Didenko N., Kunze K. Relationship between Energy Policies and Quality of Life in Russia. WIT Transactions on Ecology and the Environment, 2014, 186, 3-11
[5] Cherepovitsyn AE., Larichkin FD., Vorobiev AG., Ibrokhim A. Economical prospects of advancement in liquefied natural gas production. Gornyi Zhurnal (Mining Journal), 2018, 2. DOI: 10.17580/gzh.2018.02.09
[6] Cherepovitsyn A., Tcvetkov P. Overview of the prospects for developing a renewable energy in Russia. Proceedings of 2017 International Conference on Green Energy and Applications, 113-118. DOI: 10.1109/ICGEA.2017.7925466
[7] Gutman SS., Zaychenko LM., Rytova EV. Development strategy of Far North transport infrastructure: Problems and prospects. Proceedings of the International Conference - Education Excellence and Innovation Management through Vision 2020: From Regional Development Sustainability to Global Economic Growth, 2017, 1439-1449.
[8] Pogodaeva TV., Zhaparova DV., Rudenko DY., Skripnuk DF. Innovations and Socio-Economic Development: Problems of the Natural Resources Intensive Use Regions. Mediterranean Journal of Social Sciences, 2015, 6 (1), 129-135.
[9] Mokhatab S. et al. Handbook of liquefied natural gas, Gulf Professional Publishing, 2013
[10] Mahmood A. et al., Pakistan's overall energy potential assessment, comparison of LNG, TAPI and IPI gas projects, Renewable and sustainable energy reviews, 2014, 31, 182-193.
[11] Mak J. Configurations and methods for power generation with integrated LNG regasification, U.S. Patent No. 7,574,856, 18 Aug. 2009.
[12] Shi X., Defu C. A combined power cycle utilizing low-temperature waste heat and LNG cold energy. Energy conversion and management, 2009, 50, 567-575.
[13] Pivovarova I., Makhovikov A. Ecological regionalization methods of oil producing areas. Journal of Ecological Engineering, 2017, 18(1).

Acknowledgements

The paper is based on research carried out with the financial support of the grant of the Russian Science Foundation (Project No. 14-38-00009, The program-targeted management of the Russian Arctic zone development). Peter the Great St. Petersburg Polytechnic University.

Analysis and Recommendations of Sand Consolidation Methods to Limit Sand Production in Gas Wells

Dmitry Tananykhin, Pavel Tcvetkov, Vladislav Kamoza

Saint-Petersburg mining university, 2, 21st Line, St Petersburg 199106, Russia, pscvetkov@yandex.ru

Abstract. Nowadays more and more oil and gas fields are developed in unconsolidated sandstones. Sand production is becoming more common process and complicates the work of downhole equipment and also can lead to higher production prime cost and increase the frequency of repair works. The study of features of oil and gas fields' development in unconsolidated terrigenous reservoirs showed that in rare cases, sand production can have a positive result, but only during the operation of low-permeability reservoir presented by sandstones, and in this case sand production increases the permeability of the pay zone. Sand production in slightly cemented formation is one of the most urgent problems in many regions of Russia. Wear of equipment and its failure are due to the removal of sand and mechanical impurities. Production of the existing well reserves is also greatly reduced. Huge amounts of money are spent annually to repair wells damaged as a result of sand removal, and the average downtime of one well is about 30 days

1. Introduction

There are a lot of reservoirs which formations are composed of loose, unconsolidated formations in Russia. The most important technological issues in such hydrocarbon production from these fields are to implement sand control and prevention bridging [1].

Sand production from the reservoir is the result of unconsolidated or decaying grains of sand around the wellbore. Usually it is low or intermediate strength formation with small or no cementing material between granules. But in fact sand production is also possible from high-strength formations with good bonding of the grain. Sand production can begin immediately or it may lead to a later life cycle of the well.

Production from many wells that have penetrated these reserves, is already much longer than expected, and their further operation with a high degree of water cut can lead to the softening of the formation [2].

Oluyemi and Oyeneyin stated that the economic, operational and safety implications of sand failures require real time efficient sand management [3].

According to [4] the sand production takes place if the sand grains around the cavity is disaggregated and as the volume of sand dislodged is deposited and accumulated on production equipment continuously, cleaning will be required to allow for efficient production of the well. To restore production, the well must be shut-in, the surface equipment opened, and the sand must be manually removed. In addition the cost of the deferred production must be considered.

Work [5] describes that many operators simply choose to install sand control or stimulate the formation if it is determined to be "weak" without necessarily evaluating or predicting the sand potential. Screens and gravel packs are widely used to prevent sand flowing into the wellbore and to the surface. Nevertheless, the initial costs are generally high and they are not free from problems. Screen systems come with a risk of failure by collapsing and bursting. Gravel packs can fail because of screen failure and plugging of the high permeability gravel. Frac and pack treatments bypass the damaged area of the wellbore and also stimulate the formation, yet sanding can still occur by the production of formation sand or proppant. In a different approach, sand consolidation treatments try to achieve small or no quantity of sand production from the formation, but the treatment often results in lower regained permeability and the longevity of the treatment is not guaranteed. In many cases, even with sand control completion in place, drawdown is carefully controlled throughout the life of the well to minimize the risk of sand production. However, this can bring about a disparity between the desired and feasible fluid production rate.

Most [6] researchers both Russian and foreign agree in opinion that the currently existing ways of dealing with the removal of the destroyed particles productive formations have a number of disadvantages, in the real economic conditions of oil and gas fields development (particularly fields at the final stage of development) limits the possibility of their application.

The solution in the oil and gas production [7] is constrained due to the lack of inexpensive, highly effective means of protecting downhole equipment from mechanical particles that are brought together with the produced wells.

The main goal of the laboratory researches is the increasing of consolidation efficiency of unconsolidated productive sandstones with chemical method.

Research tasks:

1. to perform the analysis of modern technologies and technical methods used in the operation of wells draining reservoirs those are presented by unconsolidated sandstone and siltstone.
2. explore the features of the destruction of bottomhole formation zone within the Gatchina underground gas storage (UGS).
3. to develop a chemical composition designed for consolidation terrigenous reservoirs of oil and gas and investigate its basic properties.
4. to justify technology of bottom-hole zone of producing wells with use of the developed composition.

1.1. Sand control analysis of the Urengoy oil-gas-condensate field

At the Urengoy oil-gas-condensate field, the destruction of the bottom-hole zone occurs due to water occurrences that reduce the cement properties of clay rocks. Water enters the bottom-hole zone by raising the level of gas-water contact and condensing the moisture contained in the extracted gas and gas condensate [8].

The methods used in the field reduce the removal of sand, but do not solve the problem completely. To delay the overhaul, 80% of the existing wells at the Urengoy field reduce depression, but this decision is only temporary, and in the future it will be necessary to stop the wells for repairs (figure 1) [9]. For example, most of all gas wells annually exposed to sand and works with a limited yield, and 30% of the wells need to be repaired due to the formation of sand jams. In carrying out the repair mainly used solutions such as washing sand stoppers, installation of sand filter, gravel packing, cleaning bottom-hole formation zone [8].

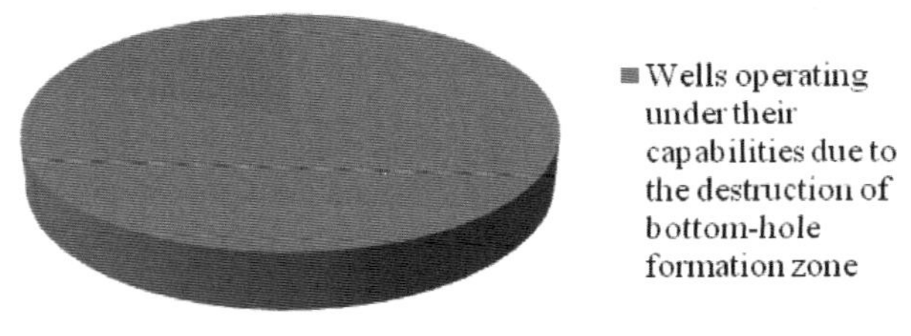

Figure 1. The proportion of wells prone by the destruction of the bottomhole formation zone at the Urengoy field

1.2. Literature and patent analysis

The analysis of the literature shows that mechanical, chemical and physico-chemical methods are used to apply to limit destruction of bottom-hole formation zone and sand production. Figure 2 shows various sand control methods in wells.

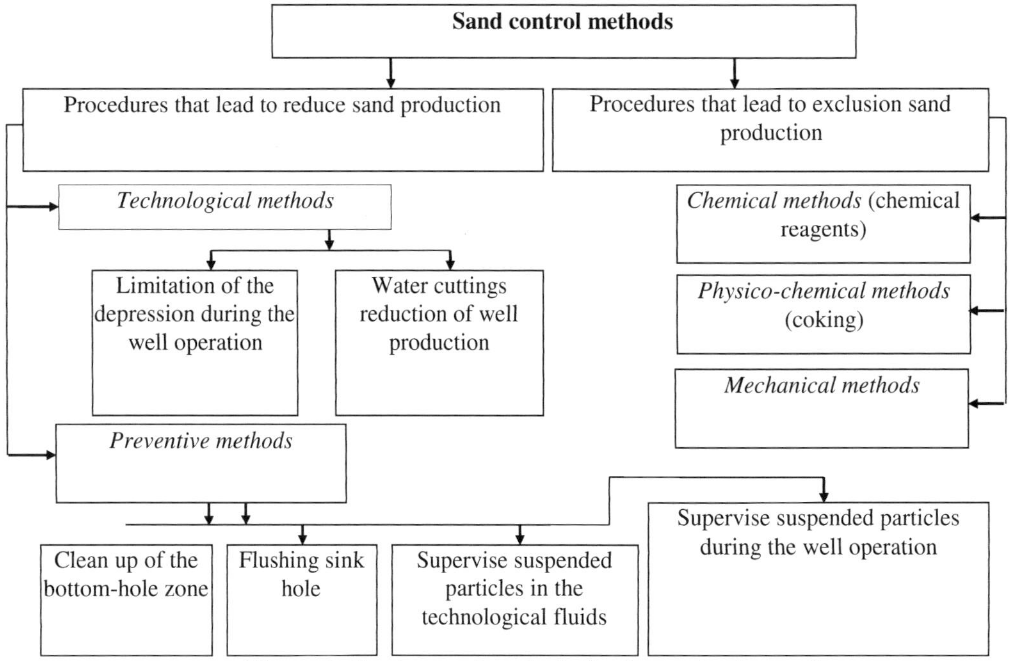

Figure 2. Sand control methods for the production wells

Mechanical methods consist of the installation the downhole filters such as wire, slatted, suspended sand, metal-ceramic etc., Such methods are generally quite effective, but the well where the sand is a result of produced water, equip filters is impractical because the problem of sand production should be addressed by the use of technologies to reduce water production.

Chemical methods based on artificial formation consolidation by astringent and cementing substances like resins cement with additives, silicates, etc. Their efficiency reveals at sufficient consolidation after treatment without significant deterioration of reservoir properties.

Physico-chemical methods include consolidation of formations by oil coking in the bottomhole zone of the reservoir, treatment with chemical reagents using thermal effect. These methods are particularly effective in the production of heavy, high-viscosity oils.

Technological methods include the constraint method of depression on the formation; water-shut-off bottom water.

2. Methodology

Laboratory experiments of the simulating of gas wells were carried out using a sand packed tube (Figure 3).

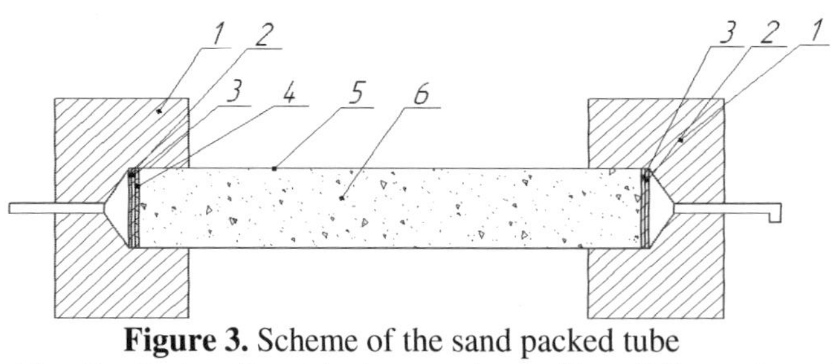

Figure 3. Scheme of the sand packed tube

1 - Collar of coreholder; 2 - O-ring; 3 - reticular membrane D_{pore}=0,2mm; 4 - reticular membrane D_{pore}=0,01mm; 5 - tube of coreholder; 6 - sand packed tube.

Disintegrated formations of Gatchina underground gas storage were used for preparing sand packed tube. Before each of the experiment formation was treated by mixture of benzene and isopropyl alcohol. Then it was dried at 100°C to stabilize of the formation mass. Depending on the desired permeability various fractions ratio from 0.1 to 0.5 mm were used.

Changes in the mass of the coreholders before and after stuffing by formations estimate to determine the porosity of the sand packed tubes. Formula 3 were used for calculating the porosity:

$$k_{пор.} = 1 - \frac{4 \cdot (m_2 - m_1)}{\rho_п \cdot \pi \cdot d^2_{вн.}}, \quad (3)$$

where: $k_{пор}$ – porosity of a sand packed tube, unit; m_2 – coreholder mass after stuffing by formations, kg; m_1 – coreholder and filters mass before stuffing by formations, kg; $\rho_п$ – grain density, kg/m^3 (in the calculation use $\rho_п = 2700$ kg/m^3); $\pi = 3{,}1415...$ – mathematical constant, $d_{вн}$ – inner diameter of the coreholder, m [10].

The investigation of reservoir properties of sand packed tube to gas was carried out according to the scheme shown in Figure 4, using the laboratory facility TKA-209.

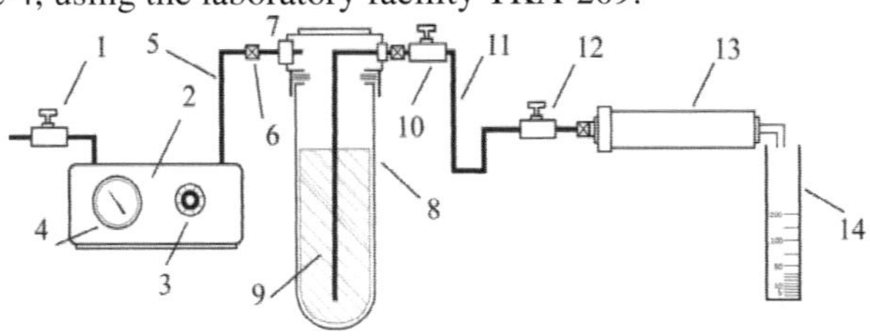

Figure 4. The scheme of the laboratory facility for simulating gas wells with using sand packed tube

1 - air valve (input); 2 – facility for pressure creation; 3 - air regulator; 4 - air pressure gauge; 5 - the air inlet tube to the reservoir fluid; 6 - tubes connecting element; 7 - drive fluid; 8 - lower fluid storage capacity (changeable); 9 - filtered liquid; 10 - valve for liquid; 11 - liquid feed tube; 12 - valve at the inlet of the coreholder; 13 – coreholder; 14 – measuring capacity.

Filtration experiments with sand packed tubes were conducted to identify the following:

- To identify the most effective concentration of the chemical components. In the experiments the range of aqueous $CaCl_2$ concentrations from 4 till 15% were observed. Concentration of $NaHCO_3$ was calculated as the reaction of the neutralization;
- Residual resistance factor R (units) - the ratio of permeability to gas of the core sample before and after injection chemical composition during the stabilization of the pressure gradient.

3. The results of the gas simulation experiments

The results of the chemical treatment of the sand packed tubes are listed in the table 1.

Table 1. The results of the chemical treatment of the sand packed tube by the aqueous solutions of $CaCl_2$ and $NaHCO_3$ under the conditions of increasing the gas flow rate

Models number	Chemical composition	Permeability to air before treatment, mkm^2	Permeability to air after treatment, mkm^2	The maximum injection air pressure gradient , MPa/m
#1	4% $CaCl_2$+$NaHCO_3$	1,14	1,01	1,32
#2	7,5% $CaCl_2$+$NaHCO_3$	1,19	1,10	1,57
#3	10% $CaCl_2$+$NaHCO_3$	1,1	1,05	1,44
#4	15% $CaCl_2$+$NaHCO_3$	1,1	0,99	1,45

Results of the filtration experiments indicate the possibility of application of the chemical composition for fixing unconsolidated productive sandstones with a slight decrease in the permeability of the bottom hole zone after chemical treatment with aqueous solutions. It is known, the main drawback of chemical consolidation methods is a significant reduction of reservoir properties. In the case of application of the developed composition minimum permeability reduction (residual resistance factor) is 1.04 times compared to the initial one. Coefficient of permeability recovery using a 7.5 % aqueous solution of calcium chloride was 93 %, as for concentration of $CaCl_2$ of 10 % – coefficient was 96 %, as for 15 % $CaCl_2$ – 91 %. Thus, the most effective concentration of the components for applications in gas wells is a chemical composition of 10 % aqueous solution of calcium chloride and 7.5 % aqueous sodium hydrogencarbonate solution.

3.1. Applications of the chemical technology on the Urengoy field

In order to improve the efficiency of chemical methods for preventing the removal of mechanical impurities and destruction of bottom-hole formation zone on the basis of St. Petersburg Mining University, a number of experiments were conducted to create a solution for fixing weakly cemented sandstones (Patent No. 2475622 "Method for fixing bottom-hole formation zone of gas wells". Modern high-precision equipment of the research laboratory of the department "Development and operation of oil and gas fields" was used for research [11].

On the basis of a number of the carried-out experiments as preventive works on the prevention of sand manifestation it is recommended to use the method of hydrophobization which consists in curing, or gel formation of hydrophobizing structures at hit in watered intervals of layer. As a result, there is fastening of weakly cemented rocks of bottom-hole zone, as well as selective tamponding of water-saturated intervals of formation, thereby limiting the inflow of reservoir water into the well, which, as is known, also contributes to the destruction of weakly cemented rocks.

During the laboratory experiments the rock was previously prepared, purifying from hydrocarbons, cementing and polluting substances by extraction with an alcohol-benzene mixture. Initial collector properties of core samples were measured at Coretest systems TBP-804 (table 2). Permeability was measured by water, so after the measurements dried each sample at low gas consumption (1,91 psi).

Table 2. Reservoir properties of core samples

Core sample name	Permeability, μm^2	Porosity
Sample #1	2.29	0.27
Sample # 2	2.1	0.25

Using sample No. 2 was pumped desiccant acetone to increase the filtration-capacitive properties of the reservoir and increase adhesion to the walls of pore channels to the hydrocarbons and treated with a water repellent, and then again dried. Made gradually purge gas of the two samples with the same exposure time and measured the loss of mass after each increase in pressure (figure 5).

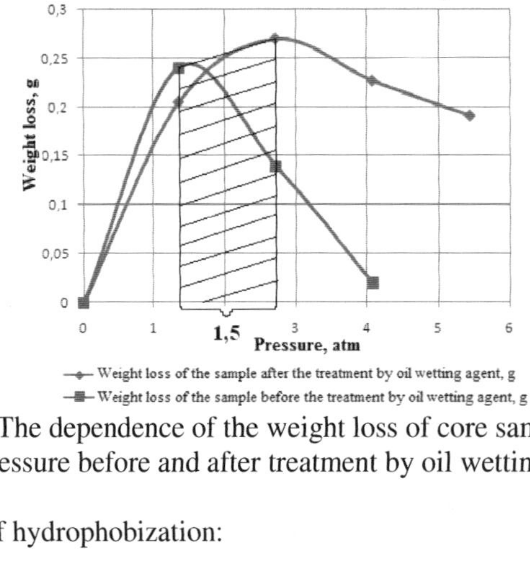

Figure 5. The dependence of the weight loss of core samples versus the pressure before and after treatment by oil wetting agent

Applications of the method of hydrophobization:

1. Consolidation of the loose formations and the prevention of sand;
2. The prevention and restriction of water influx to production wells by blocking areas of breakthrough of formation water.

If the bottom-hole zone of the well has been a violation of the structure of the reservoir rocks and, as a consequence, there is sand, we recommend the use of the chemical method of attachment weakly cemented rocks.

To achieve this goal, experiments were carried out on the injection of fastening agents into the formation. Through the sand packed tube, a consistent pumping of an aqueous solution of calcium chloride and, as a padding composition, an aqueous solution of sodium bicarbonate were carried out. As a result of the reaction, calcium ions form an insoluble compound in water, i.e., in a pore volume, a clogging precipitate is formed in the form of a fine suspension, and on the walls of pore channels in the form of solid microcrystals. Pumping of each of the specified solutions make equal portions. It is assumed that the resulting sediment in reservoir conditions will also prevent the breakthrough of formation waters by isolating the water-producing areas of the formation with a stable precipitation in water, thereby connecting to the development of stagnant and poorly drained zones of the formation.

The optimal ratio of dry matter in solutions was determined by stoichiometric calculations of the reaction with the calculation of mass fractions and laboratory studies.

After each pumping, the permeability was determined initially by gas, and in the second part of the liquid experiment.

The results of the processing of the core samples of chemical solutions is given below.

The initial permeability of the core sample for gas is K_p = 3.9 μm^2, the porosity is m = 28%.

Sequential pumping of a 10% aqueous solution of $CaCl_2$ and 13.9% aqueous NaOH solution:

$$CaCl_2 + 2\cdot NaHCO_3 = CaCO_3 + 2\cdot NaCl + H_2O + CO_2$$

Table 3. Results of sequential filtration of aqueous solutions of $CaCl_2$ and $NaHCO_3$ with successive post -increase of gas flow through the bulk model

Pressure, psi (atm)	Processing time, sec	Liquid outlet	Removal of sand
	Measurement of the gas permeability of the model: K_{pr} = 1.26 μm^2.		
5 (0.34)	90	no	no
10 (0.68)		yes	no
15 (1.02)		yes	no
20 (1.36)		yes	no
25 (1.7)		yes	no
30 (2.04)		yes	no
	Measurement of the gas permeability of the model: K_{pr} = 1.88 μm^2.		
35 (2.38)	90	yes	no
40 (2.72)		yes	no
45 (3.06)		no	no
50 (3.4)		no	no
55 (3.74)		no	no
60 (4.08)		no	no
	Measurement of the gas permeability of the model: K_{pr} = 2.67 μm^2.		
65 (4.42)	90	no	no
---		---	---
9 5 (6.46)		no	no
	Measurement of the gas permeability of the model: K_{pr} = 3.17 μm^2.		

Strengthening bottomhole with these chemical solutions for gas wells can be recommended with the following parameters of the reservoir:

1) The perforation interval should not exceed 50 m;
2) The technical condition of the well should correspond to the conditions for setting liquids under pressure into the reservoir;
3) Consistency of the permeability of the reservoir in the section, including a sufficiently high vertical permeability.

The characteristics of the chemical reagents used to fix the bottomhole zone and the technology of its utilization must meet the requirements of GOST 12.1007-76 [11].

4. Conclusion

Application of the chemical technologies for sand control of the Urengoy oil-gas-condensate field could be one of the most effective methods due to the peculiarities of the geological structure of the reserves. Currently, similar chemical technologies are not actively used, and in this regard, still large wells are subject to constant removal of sand, resulting in the formation of sand jams on the bottom of the well and equipment wear.

In the near future, authors plan to justify that the deformation of the reservoirs is determined by the ratio of its elastic properties and the properties of the saturated formation, as well as by the geometric characteristics of the deposits. To do it we are going to apply numerical calculations of simulation process.

5. References

[1] Luo W., Xu S., Torabi F. Laboratory study of sand production in unconsolidated reservoir. SPE, 2012, 158619, 1-14.
[2] Tananykhin D.S. Development of energy-efficient technologies to increase the recovery factor of unconsolidated formations, 2015, Vol. 7, No. 3.
[3] Oluyemi G.F., Oyeneyin M.B. Analytical critical drawdown failure model for real time sanding potential prediction based on Hoek and Brown failure criterion. Journal of Petroleum and Gas Engineering, 2010, 1(2), 16-25.
[4] Ikporo B., Sylvester O. Effect of sand invasion on oil well production: a case study of Garon field in the Niger Delta. The International Journal of Engineering and Science, 2015, 4(5), 64-72.
[5] Kim S.H. A predictive model for sand production in poorly consolidated sands (Doctoral dissertation), 2010
[6] Khamehchi E., Ameri O., Alizadeh A. Choosing an optimum sand control method. Egyptian Journal of Petroleum, 2015, 24(2), 193-202.
[7] Gharagheizi F., Mohammadi A.H., Arabloo M., Shokrollahi A. Prediction of sand production onset in petroleum reservoirs using a reliable classification approach. Petroleum, 2017, 3(2), 280-285.
[8] Lanchakov G.A., Moskvichev V.N., Khavkin A.I., Sorokin A.V. The use of foam systems for the reduction of sand. OOO "Gazprom dobycha Urengoy", Moscow, 2008.
[9] Chemezov P.V. Study of the influence of the destruction of bottomhole formation zone on the performance of wells (for example, Yamburg and Urengoy gas condensate fields). Krasnodar, 2004.
[10] Rogachev M., Kondrashev A. Substantiation of the technology of in-situ waterproofing in low-permeability reservoirs. Proceedings of the Mining Institute, 2016, 217 p.
[11] Tananykhin D.S., Petukhov A.V. Composition development of chemical methods of bottomhole zone treatment to limit of sand production in oil wells. National mineral resource university "Mining", 2013.

Acknowledgements

The paper is based on research carried out with the financial support of the grant of the Russian Science Foundation (Project No. 14-38-00009, The program-targeted management of the Russian Arctic zone development). Peter the Great St. Petersburg Polytechnic University.